D1806703

Scientonomy

The Challenges of Constructing a Theory
of Scientific Change

Edited by

**Hakob Barseghyan, Paul Patton,
Gregory Rupik, & Jamie Shaw**

Series in Philosophy of Science

 **VERNON PRESS**

www.vernonpress.com

In the Americas:
Vernon Press
1000 N West Street, Suite 1200
Wilmington, Delaware, 19801
United States

In the rest of the world:
Vernon Press
C/Sancti Espiritu 17,
Malaga, 29006
Spain

Series in Philosophy of Science

Library of Congress Control Number: 2021943131

ISBN: 978-1-64889-295-0

Cover design by Hakob Barseghyan.

Table of Contents

List of Figures and Tables

List of Figures

List of Tables

Contributors

Hakob Barseghyan is an Assistant Professor of history and philosophy of science at Victoria College, University of Toronto. His research interests reside at the intersection of integrated history and philosophy of science and knowledge visualization. In his *The Laws of Scientific Change* (2015), he proposed a general descriptive theory of scientific change that has since become the basis of a newly emerging empirical study of science, scientonomy. He developed a new academic workflow and implemented it by co-founding the online Encyclopedia of Scientonomy and Scientonomy journal.

Hasok Chang is the Hans Rausing Professor of history and philosophy of science at the University of Cambridge. He received his degrees from Caltech and Stanford and has taught at University College London. He is the author of *Is Water H2O? Evidence, Realism and Pluralism* (2012), and *Inventing Temperature: Measurement and Scientific Progress* (2004). He is a co-founder of the Society for Philosophy of Science in Practice (SPSP) and the Committee for Integrated History and Philosophy of Science.

Guillaume Dechauffour is a doctoral student at Sorbonne Université, Paris, France. He is associated with *Sciences, Normes, Décision* (SND) at Sorbonne Université and Cognitive Research and Enactive Design (CRED) at Université de Technologie de Compiègne. His research focuses on the naturalization of epistemology and the biocultural evolution of knowledge.

Justin Donhauser is an Assistant Professor in the Department of Philosophy at Bowling Green State University and an NSF-sponsored postdoctoral researcher at the Indiana University Bloomington. His work tackles conceptual, methodological, and normative issues at the intersection of applied sciences, and public policy and resource management. He is currently writing a book, *Robot Ethics* (under contract with Routledge).

Deivide Garcia is an Adjunct Professor at the Federal University of Recôncavo da Bahia, where he completed his Master's and PhD in Teaching, Philosophy and History of Sciences. He did his Post-doctorate at York University. He also holds a Master's degree in Logic and Philosophy of Science from the Universidad de Valladolid. His research focuses on scientific pluralism and scientific experience in Feyerabend's philosophy, which often gives support to his other interests, such as history and philosophy in evolutionary theory, Galileo's methodology, and science education.

Patrick Fraser is a graduate student in the Department of Philosophy at the University of Toronto. His research interests include logic, philosophy of physics, inter-theory relations (such as emergence and reduction), and the foundations of quantum mechanics and quantum gravity. In particular, he is interested in understanding how scientific representation and formal theorizing together yield coherent descriptions of physical reality across distinct regimes. He also studies inter-subjective and relational approaches to interpreting physical theories.

Tsung-Ren Huang is currently an Associate Professor in the Department of Psychology at National Taiwan University, Taiwan. His research interest lies at the intersection of psychology, neuroscience, and computer science. Specifically, he applies machine learning and AI techniques to discover regular patterns in the human brain, mind, and behavior, such as finding elements of attracted faces. He is also the current director of the Modeling and Informatics Laboratory at National Taiwan University.

Kye Palider has been a part of the scientonomy community since 2018 and is currently a graduate student at the University of Toronto. He has introduced epistemic reasons to the scientonomic ontology and has participated in the creation, development, and application of a diagrammatic notation for visualizing belief systems. His general interests lie in the philosophy of logic, formal epistemology, probability, pragmatism, and integrated history and philosophy of science.

Paul Patton holds a doctorate in neuroscience from the University of Chicago and has published research in perceptual neurobiology. Paul is currently a Research Fellow at Victoria College and a doctoral candidate at the Institute for History and Philosophy of Science and Technology, University of Toronto. His philosophical specialties are in the philosophy of perception, and in the philosophy of neuroscience, cognitive science, and biology, as well as in integrated HPS and scientific change. Paul is the co-editor in-chief of the online Encyclopedia of Scientonomy and an editor of Scientonomy journal.

Gregory Rupik is a doctoral candidate at the Institute for the History & Philosophy of Science and Technology at the University of Toronto. His academic research encompasses the history and philosophy of biology, exploring the remarkable similarity between the understanding of the organism which emerged from German biology at the turn of the 19th century, and the understanding of the organism central to "Situated Darwinism" in contemporary philosophy of biology. Gregory is a founding member of the Scientonomy Community and is an editor of the journal *Scientonomy*.

William Rawleigh is a graduate student at the University of Toronto's Faculty of Information. His past research has focused on the logic and semantics of science, with a particular focus on the semantic ontology inherent in the theory and practice of science as a whole enterprise. While he is still dedicated to the study of semantic ontologies, his present studies examine the intersection of semantic ontology and the social study of information with a particular eye towards how different semantic infrastructure influences the information practices of individual scientists. His own information practices are regularly disrupted by the urge to pet every dog he sees.

Ameer Sarwar is a graduate student in Neuroscience at the University of Oxford and completed his Honors BA and MA in Philosophy, both at the University of Toronto. He is interested in the intersection between biological sciences, especially neuroscience, development, and genetics, and philosophy, with a particular focus on empirical ("wet lab") research and philosophical reflection on its findings. He is also intrigued by attempts to formally model scientific phenomena and generate testable predictions.

Jamie Shaw is a SSHRC-sponsored postdoctoral fellow at the University of Toronto. He completed his PhD in 2018, under the supervision of Kathleen Okruhlik at the University of Western Ontario, with the dissertation entitled "A Pluralism Worth Having: Feyerabend's Well-Ordered Science". His primary research focuses on science funding policy and its relationship to debates on scientific rationality. This often borrows from his research on Feyerabend and pluralism. He is the editor of a collected volume on Feyerabend, entitled *Interpreting Feyerabend: Critical Essays*, and is an editor of Scientonomy journal.

David J. Stump is an Emeritus Professor of philosophy at the University of San Francisco. Educated at the University of California and Northwestern University, he is the author of *Conceptual Change and the Philosophy of Science* and co-translator of a new version of Henri Poincaré's *Science and Hypothesis*. He is also co-editor, with Peter Galison, of *The Disunity of Science*, and is author of numerous journal articles on Poincaré, Duhem, the history and philosophy of mathematics, and general philosophy of science. He is president of HOPOS, the International Society for the History of the Philosophy of Science.

Meng-Li Tsai is currently a Professor in the Department of Biomechatronic Engineering at National Ilan University, Taiwan. His research interests include neurophysiology and circulatory physiology, with a focus on decoding the somatosensory information in the brain, as well as the quantitative relationship between the activity of the autonomic nervous system and blood pressure. He also served as the editor-in-chief of Taiwan's popular science magazine *Science Monthly* and is also a novelist and poet who has published four novels and two collections of poems.

Karen Yan is currently an Associate Professor and Director in the Institute of Philosophy of Mind and Cognition at National Yang Ming Chiao Tung University, Taiwan. She employs both qualitative methods and quantitative methods to address issues in the philosophy of cognitive neuroscience, biomedical sciences, and medicine. Her current research topics include causal cognition, modeling, and operationalization in cognitive neuroscience, using quantitative tools to study the scientific changes in biomedical sciences, and the well-being measures in clinical practices.

Introduction

Gregory Rupik, Hakob Barseghyan, Paul Patton, & Jamie Shaw

University of Toronto

Over the last half-century, the contributions of historians, philosophers, sociologists, and anthropologists have produced a richer, more complex portrait of science, its past, and its transformations than was once appreciated. This more comprehensive understanding of science and its dynamics has thrown the inadequacies of past theories of scientific change, like those proposed by Fleck, Kuhn, Lakatos, and Laudan, and earlier by the logical positivists, into stark relief. Many of these theories aimed not only to account for scientific change across diverse contexts, but – by clarifying the mechanism of this change – to illuminate how science *ought* to change. While the classic theories of the logical positivists envisaged a mechanism of scientific change driven by a transhistorical scientific method, the richer portrait of science emerging from subsequent and ongoing work evinces that no single method exists. Rather, there seem to be myriad methods employed across the sciences and their histories. This presents an intimidating challenge facing anyone who attempts to construct a comprehensive theory of scientific change today: to account for the diversity and change of scientific theories as well as the diversity and change of these scientific methods.

The challenges facing any prospective theoretician of scientific change do not stop at the dynamic and diverse methods of sciences. For instance, if the means by which scientific communities assess rival scientific theories are diverse and change through time, it becomes less straightforward how the descriptive project of understanding science's changes can inform a normative project of how science *ought* to change, or if it should at all (Rupik 2019). Our growing appreciation for science's variety, social nature, and transformations task theoreticians to confront a number of claims. These include the claim that there may be nothing stable and common among the sciences or scientific changes, that there may be nothing that distinguishes scientific changes from other types of change in society, or that attempts to construct a theory of scientific change are doomed given the track-record of previous attempts (Barseghyan 2015, 81-97). Indeed, the same historical, sociological, and anthropological research that has empirically enriched our image of science can be used to render the very category of "science" so capacious as to be meaningless.

For a growing number of scholars, however, these challenges are not insuperable ones. They see an opportunity to construct a more nuanced theory of scientific change that draws lessons from the failures of past attempts, and fundamentally embraces today's empirically enriched portrait of science (Barseghyan, 2015; Scholl & Räz, 2016; Rupik, 2019). There is a commitment to integrate the history and philosophy of science in this endeavor (Integrated HPS, iHPS, &HPS; see Herring et al. (Eds.), 2019), enacting a more "naturalistic turn" wherein "*matters of fact* are as relevant to philosophical theory as they are relevant in science" (Callebaut, 1993, p. 1). It was only by considering the wealth of information now available about science's history and practice, for instance, that members of the Scientonomy Community recognized regular patterns in the change of theories, methods, and questions that would form the basis of their own theorization. While this volume will not recapitulate these historical developments here, they were a crucial impetus towards taking the idea of a general theory of scientific change seriously again.

The Scientonomy Community formed in Toronto in 2015 with the goal of refining and further developing a theory of scientific change first proposed in Hakob Barseghyan's *The Laws of Scientific Change* (2015). By setting their sights on the transformations of epistemic agents' theories, questions, and *methods*, they claim to have identified law-like patterns that appear to be common across time periods and fields of inquiry. The precise content and scope of these patterns is constantly in flux as conjectures about the laws of scientific change are evaluated on the basis of historical data and theoretical considerations. The theory aims to be descriptive, not normative, articulating the *nomic* regularities that emerge when researchers consider mosaics as their objects of study. Considering scientonomy a "science of science" has allowed the community to articulate how it has embraced the "naturalistic turn" in philosophy of science: the inseparability of theory and observation in the sciences drives the community's commitment to explicitly weave the history and philosophy of science together as observational and theoretical scientonomy, respectively (Rupik 2019).

In addition to facing the challenges peculiar to constructing a theory *of scientific change*, the Scientonomy Community has also been confronted with the more generic challenges of constructing and refining *an empirical theory*. Confronted by diverse scientific practices across time and space, scientonomers have been compelled to adjust and expand their beliefs concerning the ontology of epistemic agents, elements, and stances, modify some of the key tenets of their theory, and thus undergo their own process of scientific change. To successfully do so, the Scientonomy Community has adopted strategies similar to those used in the natural sciences: it rebooted the project of constructing a general theory of scientific change by implementing a new digital workflow. This iterative, communal workflow publicly tracks developments in scientonomy and helps to guarantee cumulative knowledge production and progress. Two pieces of this workflow worth highlighting here are its online wiki-styled *Encyclopedia*, which documents the current state of scientonomic knowledge, and its online journal *Scientonomy*, whose peer-reviewed submissions make concrete suggestions for how current scientonomic theory can be modified. One motivation for establishing this iterative workflow was the realization that any theory of scientific change is as fallible as any other empirical theory and the best strategy – implemented differently by different sciences – is to work collectively on *piecemeal* and *transparent* advancement of our communal knowledge on scientific change. Implicit in this workflow is openness to new historical evidence concerning the dynamics of theories, questions, and methods and deep respect for critique. The community's workflow has already refined scientonomic theory's ontology and taxonomy, which may serve as the foundation for a future database of epistemic communities, their mosaics, and their historical transformations – provisionally named the *Tree of Knowledge Project*. These undertakings have revealed new challenges and opportunities for a science of science.

In this communal spirit, the Scientonomy Community convoked a conference entitled *The Challenges of Constructing a Theory of Scientific Change* in 2019. Ironically, despite the fact that the ultimate ambition of the Scientonomy community is an empirically adequate general theory of scientific change, the conference did not focus directly on assessing historical episodes. Although this would obviously be essential for any viable theory of scientific change, the primary concern of the papers presented was more theoretical – that is, discerning what promises and perils belie any attempt to construct a theory of scientific change. And while the program included explicit engagements with scientonomic theory, it importantly ranged well beyond uniquely scientonomic concerns. This volume collects the proceedings of this generative, collegial gathering.

Hasok Chang's contribution notes the importance of an ontology for empirical investigations of the process of scientific change, particularly those involving the analysis of large databases. He focuses on an area of the scientonomic ontology that has so far been neglected – the ontology of scientific practice. He notes that historians of science have often offered only an imprecise analysis of the scientific activities and offers some pointers toward a rigorous analysis of the activities that epistemic agents undertake during the production of knowledge.

Kye Palider tackles the thorny issue of the nature of integrated history and philosophy of science. He summarizes the extant literature as containing two approaches: the top-down approach, where philosophy informs history, and the bottom-up approach where history informs philosophy. Palider contends that neither of these approaches can survive on their own and must be synthesized into a broader account. Palider goes on the provide a way forward which integrates and corrects the top-down and bottom-up approaches into a coherent approach based on the notion of epistemic iterations.

Hakob Barseghyan and Jamie Shaw's paper discerns ways in which normative philosophical claims about science can benefit from history. The primary worry here has been that deriving philosophical *oughts* from historical facts would commit the naturalistic fallacy. They claim that by taking the descriptive findings of scientonomy and coupling them with additional normative premises, philosophers of science can draw normative methodological conclusions which can guide future scientific practices. The paper outlines a viable path for integrated history and philosophy of science that does not relinquish normativity and avoids the problem of cherry-picking which has plagued general accounts of science.

Karen Yan, Meng-Li Tsai, and Tsung-Ren Huang's contribution explores the relationship between scientonomy and scientometrics. The authors argue that given the vast volume of scientific literature currently being produced each year, the quantitative methods of scientometric analysis are needed to test hypotheses about the process of scientific change posited by scientonomy. As a test case, they offer an analysis of the literature concerning a physiological phenomenon called heart-rate variability.

Will Rawleigh's contribution critiques the current understanding of the scientific mosaic and its constituent theories. He notes the many problems of the syntactical view of theories as sets of propositions, which is currently held within scientonomy and introduces an alternative model-theoretic view. Based on this new view, he proposes a new formulation of the third law of scientific change.

Patrick Fraser's paper is part of a bold attempt to provide a formal framework to represent scientonomic claims about theory change. Specifically, Fraser models the mosaics of scientific communities as actual instantiations of possible worlds by using a semantic approach inspired by Kripke. This provides a way forward to a possible formalization of scientonomic knowledge.

Guillaume Dechauffour positions scientonomy within a broader scientific framework of evolutionary epistemology by linking the evolution of science with evolution in general and the evolution of philosophical systems. By doing so, he aims to dissolve the problem of scientific progress in favor of much less problematic idea of scientific evolution. The paper also highlights the role of evolutionary epistemology in understanding the initial conditions of scientific change.

Paul Patton's paper explores the entities and relations within the sociotechnical domain, i.e., between epistemic agents and their tools. By reviewing the current body of scientonomic literature, Patton offers a robust new alternative to the networks of practitioners view. The paper delineates the role of individual and communal epistemic agents and epistemic tools and posits two types of relationships: the relationship of authority delegation between epistemic agents, and the relationship of tool reliance between epistemic agents and epistemic tools.

Ameer Sarwar's paper argues that the intended object of study of scientonomy lacks specificity and outlines a general conception that can help with distinguishing the object of study from background noise. It differentiates between various perspectives, types of explanations, and levels of analysis when dealing with the object of study. The paper also notes that incorporating the technical apparatus of the general system theory into scientonomy can potentially offer us a radically different way of thinking about scientific change.

Deivide Garcia's essay opens a pandora's box by discussing the many ways in which pluralism may appear or present a challenge to the currently accepted theory of scientific change within the Scientonomy Community. Garcia seeks to answer two questions in particular. First, is the current theory of scientific change necessarily exclusive, or is it possible and/or desirable to have multiple theories of scientific change? Second, is the zeroth law, which states that at a given point of time all elements of a mosaic are mutually compatible, compatible with the history of science?

In their contribution, Jamie Shaw and Justin Donhauser argue that the current scientonomic ontology doesn't quite conform with historical advances in theoretical ecology and suggests a few revisions to the scientonomic ontology. By analyzing the development of Lotka-Volterra models, broken stick models, and exergy models, they refine the scientonomic category of use by drawing a distinction between two distinct types of *use* – *epistemic* and *practical* and provide additional support for the accepted definition of *theory acceptance*.

In David J. Stump's contribution, he tackles the historically dominant issue of radical theory change. Building off of the work of others, Stump defends a conception of the relative a priori where conceptual changes can be all-encompassing. Against some, though, Stump contends these changes can be rational. While Scientonomy is traditionally neutral with regards to whether changes between theories and methods are 'rational', Stump's chapter penetrates thorny issues about the historical rationales during episodes of radical theory change.

These contributions represent an invaluable set of first steps towards fully understanding the possibility of reigniting the bold historical pictures of science of the past. That being said, there are many limitations and many other sources – both historical and philosophical – that may be fruitfully used in future discussions. For example, further engagements with recent work on cognitive attitudes, philosophy of action, or with historically oriented philosophers of science from continental traditions, Indo-China, and Latin American scholars (to name a few) will surely increase the scope of issues to be brought to bear on the topics in these pages.

Bibliography

Barseghyan, H. (2015). *The Laws of Scientific Change*. Springer.

Callebaut, W. (1993). *Taking the Naturalistic Turn, Or How Real Philosophy of Science Is Done*. The University of Chicago Press.

Herring, E., Jones, K. M., Kiprijanov, K. S., & Sellers, L. M. (Eds.) (2019). *The Past, Present, and Future of Integrated History and Philosophy of Science*. Routledge.

Rupik, G. (2019). Scientonomy: A Bold New Vision for an Integrated History and Philosophy of Science. In Herring et al. (Eds.) (2019), 19-37.

Chapter 1

The Ontology of Scientific Practice

Hasok Chang

University of Cambridge

Abstract: Any database project needs to be based on an ontology that is suitable for its subject matter. For the scientonomy project, it is important to conceive a good ontology of the human actions that constitute scientific practice, which allows rigorously conceived activity-based analyses of science. Mainstream philosophers of science have offered very precise analyses of scientific knowledge, but only in terms of beliefs and their assessments. Recent historians of science have been more attentive to the activities undertaken by scientists, but only offered imprecise analyses of them. Scientonomers have a chance to get the ontology right from this early stage of their enterprise. In this paper, I make some programmatic proposals on the ontology of scientific work, in terms of epistemic activities and systems of practice. I also offer a framework for conceptualizing epistemic agents, and some clues on the ontology of the processes of inquiry. Scientonomy as the general "science of science" should not have an overly limited ontology. In particular, I suggest that the "scientific mosaic" should include a diverse array of elements and consider other aspects of methodology in addition to theory assessment.

Keywords: scientific practice; pragmatism; epistemic activities; inquiry; knowing-how

***

1. Initial Motivations

Any database requires an ontology, and *the Tree of Knowledge* project ("a comprehensive online database of intellectual history that will trace the evolution of human knowledge") proposed in scientonomy is no exception.[1]

---

[1] https://www.scientowiki.com/Tree_of_Knowledge_Project

The Tree is important, because without the historical database, the scientonomic laws of scientific change will remain untested conjectures (or even function as tautologies), which would mean that scientonomy cannot be a truly empirical discipline (see the papers in this collection by Deivide Garcia and Kye Palider for indications of the empirical nature of scientonomic work). Curiously, this practical need for ontology is not often appreciated by philosophers. I remember being delighted some years ago to meet a computer scientist whose business card declared that he was the "Chief Ontologist" for his company – I doubt that any metaphysicians carry such a job title! The scientonomy community has been keenly aware of the need for an ontology (Barseghyan, 2018).[2] And the push for diagrammatic representation of worldviews will also make visible the necessity to have the right kinds of ontological "boxes" to use (Barseghyan, Patton, & Shaw, in press).

My own appreciation of the need for ontology comes from the philosophy of scientific practice. Analytic philosophers have mostly worked with inadequate ontologies of knowledge, the knowing agents, and their experiences. A significant part of my work in the philosophy of science is an attempt to pay more attention to scientific practices, and for this task I need to craft an ontology of activities and agents, going well beyond the familiar ontology of propositions in epistemology and the philosophy of science. In this paper, I will give some indication of that ongoing work, highlighting aspects of it that are most relevant to scientonomy. In this enterprise, it is essential to maintain connections with work in the history of science, and there is a great need for conceptual discipline there. As Hakob Barseghyan and Jamie Shaw (2022, in this collection) have argued persuasively, much work in the history of science is still framed in quotidian common-sense terms, and most philosophy of science making reference to history fails to question these terms. These terms are undisciplined – ill-defined, and more vague and ambiguous than they need to be. Again, we need a good ontology with which to understand the goings-on in scientific work.

In relation to scientonomy, what I want to offer in this paper is a friendly critique, from the perspective of someone who has been trying to think carefully about the nature of the scientific practice. My main point is that scientonomy as the general "science of science" should not have an overly limited ontology. I think the "scientific mosaic" should include aspects other than theory and methodology. And methodology should address more than assessment. Also, stable configurations deserve as much attention as change.

---

[2] https://www.scientowiki.com/Ontology_of_Scientific_Change

2. Activity-Based Analysis

A serious study of scientific practice must be concerned with what it is that we actually do in scientific work (this spirit is exemplified in Yan, Tsai, & Huang, 2022, in this collection). This requires a change of focus from propositions to activities. Percy Bridgman, the American experimental physicist who became the unwitting originator of the philosophical doctrine of "operationalism", can serve as a useful initial inspiration (see Chang, 2019, for all quotations from Bridgman and the original citations). Bridgman advocated "an attitude or point of view generated by continued practice of operational analysis. So far as any dogma is involved here at all, it is merely the conviction that it is better, because it takes us further, to analyze into doings or happenings rather than into objects or entities". And an operational analysis was only "a particular case of an analysis in terms of activities – doings or happenings", instead of analysis "in terms of objects or static abstractions", or "in terms of things or static elements". So let us begin with the recognition that all scientific work, including pure theorizing, consists of actions – physical, mental, and "paper-and-pencil" operations, to put it in Bridgman's terms. Of course, all verbal descriptions we make of scientific work must be put into propositions, but we must avoid the mistake of only paying attention to the propositional aspects of the scientific actions. That is a sure path to disconnection from practice, and it is precisely the path that analytic philosophers on the whole have taken. What I am complaining about is our habit of focusing on descriptive statements that are either products or presuppositions of scientific work, and our commitment to solving problems by investigating the logical relationships between these statements.

This way of thinking recommends an analysis of scientific practice in terms of epistemic activities (in Chang, 2011a, with further elaborations in Chang, 2014). I will state more carefully below what exactly I mean by "epistemic activity", but I begin here with an intuitive presentation. The easiest first step we can take in moving toward the habit of activity-based analysis is a grammatical one: bring back the verbs into our descriptions. Try a simple linguistic trick of taking a common noun designating a standard philosophical topic and thinking about the verb form instead. So, take "representation" and think of it as "representing", as Hacking (1983) does; take "causation" and think of it as "causing", or "making things happen" as James Woodward (2003) does, or in terms of "hunting" and "using" causes as Nancy Cartwright (2007) does. Consider how different that feels already, with fresh philosophical questions bubbling up as a result of that simple change of viewpoint. When our thinking is structured around the active verbs, a whole range of questions regarding actions emerge naturally, almost without any effort: who is doing what, why, how, and in what context?

Even theoretical work is a matter of activities, and there is such a thing as theoretical practice. What does one do when theorizing? What is the right verb for "I … the theory"? We make and assess theories, as well as use and apply them. For example, take something from science that would seem to be as far removed from activities as possible: the definition of a technical term. Instead of thinking about the nature of a definition, we can consider what one has to do in defining a scientific term: formulate formal conditions, construct physical instruments and procedures for measurement, round people up on a committee to monitor the agreed uses of the concept, and devise methods to punish people who do not adhere to the agreed uses. In one stroke, we have brought into consideration all kinds of things, ranging from operationalism to the sociology of scientific institutions.

Our basic thinking in the philosophy of science can be transformed by the habit of activity-based analysis (see Chang, 2011a for a more detailed discussion). The traditional topics of debate look significantly different when they are seen from the perspective of action. For example, confirmation can be understood better by considering various processes of hypothesis-testing. The nature of models is best discussed by considering the actions that take place in modelling. And going beyond thinking about scientific explanation in terms of logical relations between explanandum and explanans, we can consider how the act of explaining arises and how it is best performed.

3. The Ontology of Epistemic Activities and Systems of Practice

Analytic epistemology is founded on a simple ontology of knowledge: propositions, and our mental attitudes toward them. A full ontology of knowledge requires far more than that. We need a detailed and precise picture of the knowers and their epistemic activities. I will come to the epistemic agents in the next section, but let me start with their activities. So far, I have spoken about "activity" in a loose and intuitive way. In attempting to make the notion more precise, we might look hopefully to the philosophy of action, but much of the existing literature in that field is not helpful for my purposes. One clear exception is the work of Jennifer Hornsby, who has emphasized the importance of attending to activities: "Human agents participate not only in once-off actions, but also in activities" (Hornsby, 2007, p. 170). Activities occur in "a narrative which tells one in a natural way about what some person did over the course of, say, a few hours. Among the things in the narrative, there will be *activities* in which the person engaged – travelling to work, conversing with someone, composing a piece of writing, as it might be" (Hornsby, 2007, p. 169).

In my previous attempts to theorize about scientific practices, I have taken as the key unit of analysis an activity with a structure and an inherent aim. For the sake of consistency and continuity, I will begin by quoting from my best-publicized previous attempt to characterize epistemic activities, from my book *Is Water H₂O?* (Chang, 2012, pp. 15-16):

> An epistemic activity is a more-or-less coherent set of mental or physical operations that are intended to contribute to the production or improvement of knowledge in a particular way, in accordance with some discernible rules (though the rules may be unarticulated). An important part of my proposal is to keep in mind the aims that scientists are trying to achieve in each situation. The presence of an identifiable aim (even if not articulated explicitly by the actors themselves) is what distinguishes activities from mere physical happenings involving human bodies...

I will now develop this notion further. One immediate and straightforward modification is to broaden the notion of "epistemic" activities to include those activities for the evaluation and use of knowledge, as well as its production. Beyond that, there are more challenging and less straightforward developments to make. One concerns the basic ontology of an "activity". An activity is not a single act, but a complex, organized and habitual series of doings. While an individual act may be performed in a haphazard way and even result in a desirable outcome, to call something an "activity" implies a routinized and repeated doing according to a reasonably stable set of rules governing the agents' attempts to achieve the aim of the activity. And it should be stressed at the outset that there are mental activities as well as physical ones, and most of the time they are combined together.

Two particular issues deserve consideration in the ontology of epistemic activities. The first is the *composition* of an activity. In the previous definition quoted above, I indicated that an activity was "a set of mental or physical operations". I did not define what an "operation" was, but it can be used innocuously enough as a convenient term to designate the component activities into which an activity is analyzed (see Chang, 2014, pp. 74-75). More problematic is the fact that I have implied an overly atomistic picture for such analysis, as explicitly depicted by Léna Soler and Régis Catinaud's (2014, p. 82) highly cogent critique of my ideas. But if we don't want to say that an activity is made up of operations, how should we conceptualize the constitution of an activity? Following Soler, I now see an activity as a unit of coherence with many aspects to it (such as rules, beliefs, bodily motions, tools, and material objects), not as something composed of sub-units that are themselves activities. When I say "unit of coherence", the idea of "coherence" intended is the notion of "operational coherence", which indicates an aim-oriented coordination of various aspects of an activity (see Chang, in press, Section 1.4; an earlier formulation, now superseded, was published in Chang, 2017a).

The other major issue to be worked out is how aims shape activities. For a detailed understanding of actual practices, we need to think in terms of desires and values as they are instantiated in sufficiently concrete forms. Specific activities are designed to achieve proximate aims that help agents achieve their ultimate desiderata. So, in the context of the philosophy of science, there is a gap to be bridged between how we need to understand scientific practices and the ordinary way in which philosophers discuss the aims of science and the values held by scientists in highly general terms. Suppose someone asks me, while I'm striking a match, what I am trying to achieve. My answer may be "I'm trying to light a match", or "I'm trying to get a combustion-analysis of an organic compound going". Both are cogent answers, but they get at two different kinds of aim. The first answer addresses what I will call the inherent aim of the activity: getting the match to light is the whole point of the activity itself, regardless of why one is engaged in that activity – that may be to light a Bunsen burner with it, or to burn down a house, or just to watch and admire the marvelous phenomenon that combustion is. These latter reasons might be called the external functions of the activity, namely what various agents might want to achieve by means of the successful execution of that activity. The very broad aims and values just mentioned above may be regarded as external functions of various activities. An activity, as I conceive it, is partly *defined* in terms of its inherent aim, which exists regardless of any external functions that the activity may or may not serve (match-lighting is not match-lighting if one does not at least intend to light a match). Inherent aims and external functions both fall under the rubric of aims, so the talk of aims needs to be disambiguated accordingly. Some may raise the question of how we discern the aims of activities, especially when the actors involved in an activity are not articulating their aims consciously. Even in a highly self-aware realm of life like science, the actors themselves are not always reflective about their aims, and it is often left to the historians, philosophers and sociologists to engage in difficult interpretive work in order to discern the scientists' aims.

Finally, I should stress that any description of an activity we can give is a *program* of action, whether in terms of retrospective understanding or as prescriptive guidance. Such a program is bound to be abstract, in the sense of not including all the features that are present in each instance of its execution. So, any activity that we can describe is not precisely instantiated in our actual doings (while success can only be judged through how our actual doings work out). Which features to include in our description of an activity is a conventional decision, and there is no uniquely right way to identify (and classify) activities out of the stream of doings that we continually carry out in life. There is a hermeneutic dimension to activities: the operational coherence of an activity is about pragmatic sense-making.

Epistemic activities normally do not, and should not, occur in isolation. Rather, each one tends to be practiced in relation to others. Often they form a network that is dense enough and large enough to deserve to be called a "system of practice". To refer back to my 2012 publication (Chang, 2012, p. 16; see also Chang, 2014, p. 72):

> A system of practice is formed by a coherent set of epistemic activities performed with a view to achieve certain aims... Similarly as with the coherence of each activity, it is the overall aims of a system of practice that define what it means for the system to be coherent.

I now want to make a more careful consideration of how epistemic activities come together coherently to constitute a system of practice. Though I have made some historical studies to show how systems of practice formed in various specific situations, I have not previously provided a well-developed general characterization of the process. This is similar to what was lacking in Kuhn's discussion of a paradigm as a "disciplinary matrix", in which he described various types of elements constituting a paradigm but did not specify how exactly they came together (Kuhn, 1970, pp. 181-187). The various activities in a system are not merely juxtaposed to each other, or merely practiced simultaneously by the same actors; rather, they are brought together in very particular ways that are designed to meet certain aims.

Let me illustrate some basic intuitions with the help of an example, staying within the scientific context. Antoine Lavoisier created a new system of practice in chemistry, whose main epistemic activities included: making various chemical reactions, collecting the reaction products, especially gases, identifying various substances through standard chemical tests, analyzing organic substances by combustion, measuring the weights of the ingredients and products of reactions, tracking chemical substances through those weight-measurements, and so on. Some of these were well-established activities in chemistry, and others were more novel. These activities co-existed and worked together within the system. But what exactly are the relations between such activities that come together to form a system of practice?

Like activities, systems of practice also have aims. But one key difference is that a system of practice does not have a single inherent aim. In fact, that may be considered the chief difference between an activity and a system. System-wide aims may be easier to discern in the context of applied science, but they are clearly present in systems of pure science, too. To continue with the example just given, we may ask what the overall aims of the Lavoisierian system of chemistry were. This is a subtle historical question, but the more obvious aims would include attaining the knowledge of the composition of various substances, explanations of chemical reactions in terms of compositions, and a taxonomy of substances according to their compositions (this was a "compositionist" system of chemistry, as discussed in detail in Chang, 2012, sec. 1.2.3, and Chang, 2011b). There is both dependence and independence among such aims. So, knowing the compositions was essential for the other two aims of explanation and classification, because in this system there was a commitment to make explanations and classifications on the basis of compositions. On the other hand, classification and explanation are largely independent aims, and one could be pursued without the other at least to an extent.

To take a different kind of example: consider the game of soccer – not an individual match, but the whole institution of it. This may be considered a system of practice. It does not have a unitary inherent purpose, so it is not an activity in the strict sense as I define it here. This is not to deny that some particular activities *within* soccer have unitary inherent purposes: for example, the inherent purpose of goal-keeping is to prevent the other team from scoring, and the inherent aim of passing is to give the ball to a player of one's own team. But isn't winning the inherent aim of the game itself? That may be said to be the aim of each team, but not of the whole system, to which winning does not apply. If we ask seriously about the aim of the whole system of soccer, there is no easy answer, but the answer will be multiple: to provide entertainment for the people, to provide income and profits to some individuals and entities, to promote health and fitness in society, to contribute to community solidarity, etc. All these aims come into play, separately or together, as people and groups work to maintain and improve the workings of the system.

The coherence of a system of practice consists in an effective coordination of the external functions of various activities for the achievement of system-level aims, which go beyond the inherent purposes of the individual activities that are pulled together to constitute the system. A system of practice is something *put together* by agents, and maintained by specific effort; it only exists because someone upholds it by means of some mechanisms for propagation and maintenance.

4. The Ontology of Epistemic Agents

If we want to think seriously about knowledge in the context of action, we cannot neglect the actor (see Patton, 2022 in this collection for some careful consideration of epistemic agents). Traditional discussions in epistemology or even the philosophy of science present the scientist only as a ghostly being that either believes or doesn't believe certain descriptive statements, fixing his beliefs following some rules of rational thinking that remove any need for real judgment. Whatever does not fit easily into this bizarre and impoverished picture, we tend to denigrate as matters of "mere" psychology or sociology. We need a more serious understanding of the scientist as a real agent, not as a passive receiver of information or an algorithmic processor of propositions. I cannot hope to do justice to all the relevant matters of psychology, sociology and biology involved here, but at least I want to express a recognition of what we need to be able to think about competently, and personal humility in not knowing how to do it well.

So let's begin with the motto from William James and Marjorie Grene: the knower is an actor. The first step is to recognize purposive behavior in the knower, at least in terms of instrumental rationality. The most basic thing about an agent is that she acts with purpose, striving to achieve certain aims that are formulated on the basis of her desires and beliefs: the agent takes the kind of actions that she believes will contribute towards the satisfaction of her desires. That is the "standard story of action" in the philosophy of action, as Hornsby calls it, according to which actions are "belief-desire caused bodily movements" (2007, p. 180, also p. 165). Hornsby has criticized this account strongly as not giving a truly active role to the agent: the standard story "leaves agents out… and, in any of its versions, the standard story is not a story of agency at all" (2004, p. 2). But actually, in philosophy of science even a proper recognition that the epistemic agent has desires, rather than just beliefs, would be a significant advance. There are many types of pleasures that motivate scientists (and human beings in general), including physical comfort and sensual well-being, abstract and concrete understanding, love and conviviality, self-esteem, security, legacy, and a sense of beauty, order and coherence. The end of this list overlaps with what has been recognized as "epistemic values", by some philosophers of science. And on top of that, of course, we have to think about the things that various people have regarded as "the aim(s) of science": truth, empirical adequacy, economy of thought, etc. In order to understand concrete practices, we need to see how such general aims and desires shape the specific proximate aims that drive particular epistemic activities.

Aside from desires and aims, epistemic agents have beliefs indeed. And we need to think more broadly than just looking at beliefs under explicit consideration. There are also things we take for granted without examination, and such presumptions are necessary for enabling any kind of actions. Of particular importance are expectations concerning the future. Expectations are often not beliefs at all, if by belief we mean a conscious assent to an articulated proposition. They can even consist in *not entertaining* certain possibilities. When I am walking along as normal, my expectations involved in that activity will not be exposed or even formulated until they are met with something incoherent with them, such as the tremors of an earthquake, or the left arm grabbed by an excited old friend, or a gaping hole in the pavement. Scientific practice is also full of expectations, sometimes guiding our activities smoothly, sometimes preventing certain activities, sometimes making us attempt something repeatedly without a clear sense of why.

Something else often neglected in philosophical accounts of science is epistemic agents' capabilities (or capacities), which are discussed mostly in the discussions of ethics and human flourishing. Hornsby criticizes the standard story of action for leaving out agents' capacities, too (2007, p. 170; see also 2004, pp. 20-22):

> When abilities are allowed a place in the explanation of action, it becomes clear how narrowly focused are the explanations from agents' reasons given in the standard story. I said that it would be a sort of magic if someone's intentionally doing something were consequential merely on their having a desire and a belief. And of course we know that reason-explanations succeed only on the assumption that agents are possessed of various capacities. Some such capacities, such as that which human agents arguably have for self-determining choice, are extremely generic in their scope... Abilities are explanatory states which operate at an intermediate level of generality as it were. So too, as we shall see now, do some states of knowledge.

It is important to recognize both mental and physical capabilities here, and also their mutual entwinement. We should also remember Michael Polanyi's (1958, ch. 4) emphasis on the role of skills in scientific work, which paid due attention to the embodiment of knowledge. Now, if specifically questioned, no one would deny that epistemic agents have capabilities, but we lose sight of some important things by not keeping them explicitly in mind. Perhaps most importantly, considerations of scientific rationality are greatly hampered if we do not consider what the capabilities of the actors involved are, because the judgment of what is rational for them to do (or attempt) depends greatly on what they are in fact able to do (remember the old lesson in ethics: "ought implies can"). Different people have different capabilities, and most specific capabilities have to be learned, so we need to consider the process of learning and training. The consideration of capabilities should be clearly present in the discussion of a wide range of philosophical issues, such as observability, testability, simplicity, and incommensurability – and therefore also realism, demarcation, confirmation, theory-choice, and so on.

Epistemic agents also make choices and judgments. The standard story of action critiqued by Hornsby treats beliefs and actions as objects of causal explanation, which makes it difficult to allow judgement on the part of the agent. That is the same regardless of the type of cause that the explanations appeal: utility-seeking, cognitive-psychological, neurological, sociological, what have you. For example, in an interest-based sociological explanatory schema, the picture of the individual is actually not that different from, and just as impoverished as, that of the utility-maximizer in the individualist rational-choice theories. Instead, we need to find ways of giving some real meaning to words like "choice" in the phrase "theory choice" and "decision" in "decision theory". I will not pretend to have a solution to the problem of free will, but the least we can do is to acknowledge that even if the correct account of actions must be ultimately deterministic, decisions to act in specific ways are determined by a distinct combination of aims, beliefs and capabilities for each individual and each community that takes the actions.

So far, I have spoken of epistemic agents as if they were isolated individuals. But it is crucial to take into account is their social nature (see Patton, 2022, in this collection on the priority of the social in epistemology). Even if we do not consider truly collective agents, it is undeniable that individual agents only exist and function in a social setting, which means that there are irreducibly social dimensions of knowledge. This is no place to attempt a full social ontology, social epistemology or sociology of knowledge, but I do want to offer some reflections about the relation between the individual and social that are both novel and pertinent. This recognition of the social dimension should not degenerate into the unhelpful slogan that "everything is social". For something like scientific knowledge to arise, we must have independent individuals, as well as unindividuated society. And the society–individual interaction needs to be conceived in an iterative way, avoiding reductionism in both directions. Individual action and cognition are grounded in society, but they are not "merely" social, and we should not presume that they are explainable or even fully describable by means of community-level factors alone. The rational individual agent arises from the social matrix, but with a capacity for independent thought and dissent. And from the association of such individuals emerges a higher level of sociality that forms an integral aspect of life as we know it, and also forms the basis of fully developed systems of knowledge, especially scientific knowledge. This iterative formation of the individual from the social and the social from the individual continues indefinitely.

5. Conceptualizing the inquiry process

Having considered the ontology of epistemic activities and agents, let me now turn my attention more specifically to the processes of *inquiry*. I think that should be a central concern for scientonomy with its focus on scientific change, since inquiry is the vehicle through which scientific change happens. In my thinking about inquiry I draw a great deal of inspiration from the classical pragmatists. The emphasis on the dynamic and active processes of inquiry was already clear in C. S. Peirce's pioneering work, starting with his early classic "The Fixation of Belief" published in 1877, where he says that inquiry is the "struggle to free ourselves" from a state of doubt and "pass into the state of belief" (Peirce, 1877, p. 5). In Cheryl Misak's reading (2013, pp. 32-33): "In Peirce's view, what is wrong with the state of doubt is not that it is uncomfortable, although it is in fact uncomfortable. What is wrong with doubt is that it leads to a paralysis of action" (see Dechauffour, 2022, in this collection on the scientonomic treatment of problems).

However, much as Peirce's emphasis on the context of action set the pragmatists in a productive new direction, in another way he was a traditionalist. While Peirce emphasized the process of inquiry, what he focused on was the process of fixing *beliefs*, as with today's epistemologists whose point of departure is the conception of knowledge as justified true *belief*. For Peirce "the sole aim of inquiry" is to settle belief, and a permanently settled belief is a true belief. Talking about inquiry as solely focused on beliefs does two unhelpful things. First of all, it consolidates the assumption that knowledge is merely propositional. Secondly, it implies that belief is the only relevant epistemic attitude that we need to be concerned with. Belief may be the only epistemic attitude that ultimately matters in religious life, but that is not the case in other walks of life, including science. In analytic epistemology, we have become too used to talking of "belief" simply meaning an acceptance of a proposition as true, and insensitive to the other everyday connotations of the term. As a reminder, here is the definition of "belief" in the Google dictionary: "1. an acceptance that something exists or is true, especially one without proof. 2. trust, faith, or confidence in (someone or something)". Neither definition actually gives a sense of "knowledge". In fact, the first definition implies a direct contradiction between belief and justification: if you *know* something, why would you need to have *belief*? Of course, philosophers are free to use words technically, even in ways that do violence to their everyday meanings and connotations. But this little reflection on the meaning of "belief" is an opening to a wider issue: epistemic activities are driven by a variety of different epistemic attitudes, including acceptance and pursuit.

"Belief" is actually a much too vague and multifarious notion. Sometimes belief is a hard-wired compulsion, sometimes it comes by choice, and sometimes it arises as a consequence of other choices we make. There is belief, make-believe, and suspension of (dis)belief. There is articulated affirmation, unthinking presumption, and complicit social acceptance. There is also religious and quasi-religious faith that is unshakable and absolutely certain. Belief as intended by philosophers sometimes points to *trust*, which is a separate issue from *affirmation*, which is made as a result of explicit assessment. And when we have neither trust nor affirmation (nor its opposite, *negation*), we have the attitude of *doubt*. Inquiry should be judiciously guided by the various epistemic attitudes that we have toward various propositions, and it is important to distinguish *epistemic actions* from epistemic attitudes. *Accepting* a theory or a proposition is an epistemic action (generally based on trust or affirmation), so the contrast often drawn between belief and acceptance is not quite apt. Acceptance, as such, is a minimal and uninteresting kind of action ("I accept the general theory of relativity" – what of it?). In the context of inquiry, rather than acceptance as such, the *use* and *pursuit* of a theory or proposition are two important kinds of actions in inquiry. Use may seem straightforward, but there is actually a very interesting complication arising from the fact that we often appear to use theories or models that we negate. The troubling appearance here is deceptive, and can be resolved into a few different untroublesome scenarios. Pursuit is what we do with an idea that we want to develop further. It is an action appropriate for propositions that are in doubt, but there is also much pursuit of affirmed propositions made in a bid to extend their truth.

It is also helpful to recognize that there are very different kinds of processes that fall under the rubric of "inquiry". Treating inquiry as a homogeneous kind of thing would be a mistake. Kuhn (1970) was keenly aware of this issue when he drew a distinction between "normal" and "extraordinary" research, the former category being "research under a paradigm" and the latter being what scientists do when they feel that it is necessary to depart from the ruling paradigm in order to solve an urgent problem. Two important basic types of inquiry can be found in the realm of Kuhnian normal science. First, there is an inquiry that is a straightforward kind of fact-gathering, filling in the blanks laid out by what we already know and think. For example, the naturalist counts samples of various insects and birds on an island to estimate the population sizes each year. The chemist employs the tried-and-tested analytical methods in order to identify a substance as one of the well-known chemicals. The forensic pathologist needs to determine the cause of someone's death, and chooses from an acknowledged array of possibilities, excluding supernatural causes, spontaneous human combustion, etc. In such cases, the relevant concepts are already well-defined, and the necessary methods of investigation are also well-established before the particular inquiry begins. It is not that fact-gathering is always straightforward. Consider, for example, the famous story of the race to "crack" the structure of the DNA molecule; the double-helix structure was not anything seen before in a molecule, yet once it was revealed it was nothing that changed people's understanding of how organic molecules could be formed. So in the realm of biochemistry what was gained was no more than a delightful new fact (while in the realm of genetics it was a field-changing discovery).

Contrast fact-gathering to what I propose to call "learning-how-to", in conscious parallel to Gilbert Ryle's (1946) concept of "knowing-how" (or, knowing-how-to, to be clearer). This is the kind of inquiry aimed at a pre-ordained outcome. In fact-gathering the kind of answer is pre-determined, but the specific answer is unknown. In learning-how-to, the final destination is pre-determined, but we don't know how to get there. The "how" is the object of inquiry here, and what is achieved if we are successful is a specific type of ability (though some facts might also be learned along the way). Take something like human cloning, which is a great challenge even though the desired outcome is quite well-defined. Very simple phenomena can pose the same sort of challenge, too – for example, getting a fusion chain reaction in hydrogen was a great technical challenge, overcome relatively soon for bomb-making in the 1950s (and still quite unsolved for making an energy source that can be harnessed). Learning-how-to is not confined to the business of finding engineering solutions. Basic science is also full of learning-how-to challenges. Consider how Newton had to invent the calculus in order to learn how to compute the trajectories of planets, even after he had figured out the laws of motion and gravitation.

Both fact-gathering and learning-how-to are well-controlled kinds of inquiry. But perhaps the most exciting and significant type of inquiry begins with neither the final outcome nor the methodology is pre-determined, which I will call "unrestrictive research". Kuhn described this sort of inquiry vividly with his notion of extraordinary research that often brings about revolutionary change, but it can take place at a much smaller scale, too (as Kuhn did recognize with examples like the discovery of X-rays). I draw my main inspiration on this subject from the classical pragmatists, and in my view, the consideration of the inquiry process was a distinctive contribution from pragmatism unmatched by other philosophical traditions. I have already noted Peirce's view that inquiry begins with a disturbed and unsettled state of doubt. Dewey developed this perspective to the full: "Inquiry is the controlled or directed transformation of an indeterminate situation into one that is so determinate in its constituent distinctions and relations as to convert the elements of the original situation into a unified whole" (Dewey, 1938, pp. 104-105). Dewey's view was clearly action-oriented, seeing inquiry as an activity of an organism in its environment. Dewey was more explicit than Peirce about seeing inquiry as a process.

Dewey's view is promising, but also somewhat vague. There are significant questions to be addressed, both about the aims and the processes of inquiry. When we take Dewey's view of inquiry seriously, one thing that becomes clear is that it is not a simple matter to say what the aim or product of inquiry is. What is Dewey trying to get at when he says that inquiry aims at converting the elements of the problematic situation into "a unified whole"? We know that Dewey did not hold a reductionist vision of theoretical unity as the ideal of knowledge. Rather, I think he was pointing to something like operational coherence. He also mentions the examination and development of meaning in the last stages of inquiry. So some sort of *sense-making* is clearly being proposed as an important aim of inquiry. Inquiry beings from puzzlement, tension or unhappiness, and it is driven by the desire for understanding, or at least the desire to escape the discomfort of not understanding what is going on.

We really need to understand the workings of unrestrictive research, both because it is the most creative kind of intellectual work that we should be celebrating as the height of human achievement, and because this kind of achievement is actually the foundation of all cognitive activity, including the other kinds of inquiry discussed just above. Successful outcomes of unrestrictive research lie at the foundation of language use, mathematics, experimental design, causal reasoning, theoretical explanations, and almost all other aspects of intelligent life. If we take Dewey's view of inquiry seriously, it becomes clear why what I have called learning-how-to and fact-gathering above are really just unusually simple sub-types of unrestrictive research.

6. Concluding Remarks

What are the implications of the ideas presented above for the practice of scientonomy? I will take the currently accepted ontology in scientonomy to be what is expressed in the entry on the "Ontology of Scientific Change" in the *Encyclopedia of Scientonomy*, which asks: "What are the fundamenttal *entities*, *processes*, and *relations* of scientific change?", a question first formulated by Hakob Barseghyan in 2015. The following list of currently accepted best theories on the subject gives a good indication of the types of elements recognized within scientonomic ontology:[3]

- Epistemic Agents - Community (Barseghyan-2015)

- Epistemic Elements - Questions and Theories (Barseghyan-2018)

- Epistemic Stances Towards Questions - Acceptance (Rawleigh-2018)

- Epistemic Stances Towards Methods - Employment (Barseghyan-2015)

---

[3] https://www.scientowiki.com/Ontology_of_Scientific_Change

- Epistemic Stances Towards Theories - Acceptance, Use and Pursuit (Barseghyan-2015)

- Theory Assessment Outcomes (Patton-Overgaard-Barseghyan-2017)

The accepted theory of epistemic elements (Barseghyan, 2018) states that "the two basic classes of elements that can undergo scientific change are questions and theories, where each theory is an attempt to answer a certain question, and method is a subtype of normative theory". These elements make up the scientific mosaic, which are borne by epistemic agents. Some significant changes and ways forward are proposed in subsequent articles published in *Scientonomy* journal (www.scientojournal.com), but the points I make below should be considered questions and suggestions in relation to the currently accepted existing ontology in scientonomy.

The most significant recommendation I would make is that the ontology should be broadened out. The scientific mosaic ought to include other items in addition to theory, question, and methodology (see Rawleigh, 2022, in this collection on reconceiving the mosaic, as well as Barseghyan, 2018). Specifically, I recommend that it should include types of actions, and other items relating explicitly to actions and agents (e.g., values and aims, experimental techniques, capabilities, experience, etc.). All these can be very important things that undergo change in the process of scientific change. Also, I would urge that individuals, as well as communities, be admitted as epistemic agents. This would allow scientonomy to capture important interactions that happen between individuals, and between an individual and a community. Generally, I would offer for scientonomers' consideration the ideas I have outlined above about epistemic activities and systems of practice, epistemic agents and their epistemic attitudes and actions, and different types of inquiry.

Even if no entirely new types of entities are added to the ontology, the way that existing types of entities are conceived can be updated and enriched. For example, the discussion of methodology should address more than methods for the assessment of theories, because there are methods for generating and using knowledge as well. Methodology should address not only criteria for theory-assessment, but strategies of theoretical and material discovery and invention, as well as measurement and manufacture. Especially given scientonomy's focus on scientific change, it is odd that scientonomy seems to neglect the process of discovery. And any engagement with the literature in the history of science will demand an ontology into which discovery can be naturally accommodated. Discovery remains a key narrative feature in the history of science, and much attention has been paid to it by historians, scientists and the general public, though not by philosophers. From the perspective of activity-based analysis, we must again ask what one does when one discovers something. There will be a variety of types of activities involved, including observation, conceptualization, confirmation and persuasion.

Even the conceptualization of theories can benefit greatly from a serious consideration of activities and agents. Scientonomy seems to have an unhelpfully traditional tendency, focusing too much on theories and treating them as sets of propositions. A simple and beneficial relaxation would be to include concepts and individual propositions, as well as whole theories, in the scientific mosaic. Going beyond that, I think it would be important to recognize that theories are objects of various epistemic actions (acceptance, use and pursuit should be taken as actions, not stances). This angle should be fully adopted, always asking what the scientists are doing in relation to theories.

Finally, there is a point that is perhaps too fundamental to be helpful in relation to the basic aims of scientonomy: stable configurations in science deserve as much attention as change. Scientific practice is not all about change. Stable configurations can be as important as change, and the labor of maintaining stability should be understood along with mechanisms of change (the first law is not quite enough for this).

Hopefully, the issues that I have raised here can become *accepted* questions in scientonomy. And I do not mean to suggest that the actual work of scientonomy should wait until all the questions of ontology are settled, in which case nothing will ever get done. Rather, I would suggest that an adaptable working ontology should be adopted, with an expectation that adjustments are inevitable and can be desirable. This is the art that, for example, those who manage the Diagnostic and Statistical Manual (DSM) of mental-illness classification have had to learn (Chang, 2017b). One productive method of proceeding would be to plunge into the pilot projects on history, which would give very useful clues on how well the working ontology applies and where adjustments are needed.

Bibliography

Barseghyan, H. (2015). *The Laws of Scientific Change*. Springer.

Barseghyan, H. (2018). Redrafting the Ontology of Scientific Change. *Scientonomy*, 2, 13-38.

Barseghyan, H. & Shaw, J. (2022). Integrating HPS: What's in it for a Philosopher of Science? In this volume, 41-65.

Barseghyan, H., Patton, P., & Shaw, J. (Eds.) (in press). *Visualizing Worldviews: A Diagrammatic Notation for Belief Systems*.

Cartwright, N. (2007). *Hunting Causes and Using Them: Approaches in Philosophy and Economics*. Cambridge University Press.

Chang, H. (2011a). The Philosophical Grammar of Scientific Practice. *International Studies in the Philosophy of Science*, 25, 205-221.

Chang, H. (2011b). Compositionism as a Dominant Way of Knowing in Modern Chemistry. *History of Science*, 49, 247-268.

Chang, H. (2012). *Is Water H$_2$O? Evidence, Realism and Pluralism.* Springer.

Chang, H. (2014). Epistemic Activities and Systems of Practice: Units of Analysis in Philosophy of Science after the Practice Turn. In Soler et al. (Eds.) (2014), 67-79.

Chang, H. (2017a). Operational Coherence as the Source of Truth. *Proceedings of the Aristotelian Society*, 117, 103-122.

Chang, H. (2017b). Epistemic Iteration and Natural Kinds: Realism and Pluralism in Taxonomy. In Kendler & Parnas (Eds.) (2017), 229-245.

Chang, H. (2019). Operationalism. In Zalta, E. N. (Ed.) (2019). *The Stanford Encyclopedia of Philosophy (Winter 2019 Edition).* Retrieved from: https://plato.stanford.edu/archives/win2019/entries/operationalism/.

Chang, H. (in press). *Realism for Realistic People.* Cambridge University Press.

Dechauffour, G. (2022). Thinking Big: The Science of Change and the Historicity of Scientific Method. In this volume, 123-142.

Dewey, J. (1938). *Logic: The Theory of Inquiry.* Holt, Reinhardt & Winston.

Hacking, I. (1983). *Representing and Intervening.* Cambridge University Press.

Hornsby, J. (2004). Agency and Actions. In Hyman & Steward (Eds.) (2004), 1-23.

Hornsby, J. (2007). Knowledge and Abilities in Action. In Kanzian & Runggaldier (Eds.) (2007), 165-180.

Hyman, J. & Steward, H. (Eds.) (2004). *Agency and Action.* Cambridge University Press.

Kanzian, C. & Runggaldier, E. (Eds.) (2007). *Cultures: Conflict – Analysis – Dialogue.* De Gruyter.

Kendler, K. S. & Parnas, J. (Eds.) (2017). *Philosophical Issues in Psychiatry IV: Classification of Psychiatric Illnesses.* Oxford University Press.

Kuhn, T. S. (1970). *The Structure of Scientific Revolutions*, 2nd ed. University of Chicago Press.

Misak, C. (2013). *The American Pragmatists.* Oxford University Press.

Patton, P. (2022). Scientonomy and the Sociotechnical Domain. In this volume, 143-175.

Patton, P., Overgaard, N., & Barseghyan, H. (2017). Reformulating the Second Law. *Scientonomy*, 1, 29-39.

Peirce, C. S. (1877). The Fixation of Belief. *Popular Science Monthly*, 12, 1-15.

Polanyi, M. (1958). *Personal Knowledge: Towards a Post-Critical Philosophy.* University of Chicago Press.

Rawleigh, W. (2018). The Status of Questions in the Ontology of Scientific Change. *Scientonomy*, 2, 1-12.

Rawleigh, W. (2022). Reconceiving Scientific Mosaics: A New Formalization for Theoretical Scientonomy. In this volume, 83-103.

Ryle, G. (1946). Knowing How and Knowing That: The Presidential Address. *Proceedings of the Aristotelian Society, New Series*, 46, 1-16.

Soler, L. & Catinaud, R. (2014). Toward a Framework for the Analysis of Scientific Practices. In Soler et al. (Eds.) (2014), 80-92.

Soler, L., Zwart, S., Lynch, M., & Israel-Jost, V. (Eds.) (2014). *Science after the Practice Turn in the Philosophy, History and Social Studies of Science.* Routledge.

Woodward, J. (2003). *Making Things Happen: A Theory of Causal Explanation.* Oxford University Press.

Yan, K., Tsai, M.-L., & Huang, T.-R. (2022). Integrating Scientonomy with Scientometrics. In this volume, 67-82.

Chapter 2

Ways of Integrating HPS: Top-down, Bottom-up, and Iterations

Kye Palider

University of Toronto

Abstract: Philosophy of science and history of science have been unable to integrate in a meaningful fashion. The major difficulty has been the question of how the history of science can inform the philosophy of science. By making several distinctions to characterize the type of philosophy of science relevant for integrated HPS, I show how traditional approaches to integration failed. These include a top-down and a bottom-up philosophical approach to integrated HPS. I then present a more fruitful way of integrating the disciplines, that of iterations.

Keywords: integrated history and philosophy of science; inductivism; hypothetico-deductivism; epistemic iterations

***

1. Introduction

Philosophy of science (PS) and history of science (HS) have been put at odds with one another whenever their arranged marriage has been put in question. Yet, there have been many advocating for an intimate relation between the two disciplines in the form of integrated history and philosophy of science (&HPS), and many times this has resulted in failure. The two primary questions to ask in this interdisciplinary field are: What does the *history* of science have to offer the philosophy of science? What does the *philosophy* of science have to offer the history of science? The latter question has often had many answers and is not nearly as problematic as the former question. Many answers to the former question have been given, and theories of how the history of science can improve our understanding of science, even in the modern day, have often been unsuccessful. It is essential that both questions have answers to form a mutually beneficial cross-disciplinary study of science through historical and philosophical avenues. I hope to review some of the failures in integrating the history and philosophy of science and shed light on a promising way of integrating history and philosophy of science – iterations.

2. Characterizing Philosophy of Science for &HPS

One major issue that has plagued &HPS has been the ill-defined character of PS. Arguments against the possibility of &HPS have often tacitly assumed certain characterizations of PS which have led to conclusions that only apply to these specific characterizations.[1] The sentiment that there is a need for a proper characterization of PS has also been noticed by other authors, but no immediate solution has been presented (Schickore, 2011a, p. 469; also, Richardson, 1992, p. 42). In this section, I set out to characterize PS for &HPS by making several distinctions: first a metaphysics/epistemology distinction, then a PS/HS relevant concepts distinction within epistemology, and a normative/descriptive distinction within &HPS concepts. Lastly, I present some examples.

The initial distinction to be made is between the metaphysical and the epistemological sides of PS (this topic is also covered in Barseghyan & Shaw, 2022, in this collection). &HPS will in general be more interested in the epistemological side of things, but simply distinguishing between metaphysics and epistemology is insufficient to properly characterize the type of PS most important to &HPS. On the metaphysical end, we generally have PS intermingling with scientific ontological concepts, their definitions and related theories. These include debates about realism and anti-realism, theories of causation, the nature of space and time, evidence, error, replication, observation, etc. These are to be contrasted with the more epistemological PS that more closely considers knowledge and its acquisition. Questions are covered on acceptance, theory choice, justification, discovery, etc. However, these two types of PS are not disjoint. Many metaphysical questions interact with epistemological ones. For example, the topic of evidence is relevant to both metaphysics and epistemology, and as such, cannot be placed exclusively under the banner of either of them. &HPS does however seem to be primarily

---

[1] See for instance Nickles, 1986, p. 255. Here Nickles argues that if there is no fixed normative methodology of science, there can be no general theory of scientific knowledge. This is conflating the normative and descriptive theories of scientific knowledge. Although a normative one may be an untenable enterprise, perhaps a descriptive theory of scientific knowledge can be achieved.

concerned with epistemology, and how science is practiced, but one needs to draw further insight into which concepts are relevant for &HPS.

Historians of science for the most part do not concern themselves with many debates in the philosophy of science in their writings. There are many philosophical concepts that have no immediate value for HS: error, replication, realism, and probability, to list a few. This does not mean that HS is not interested in any PS concepts. Here I aim to outline but a few of these mutually relevant concepts between HS and PS. For instance, the question of demarcation is important both for HS and PS. What is science? Without at least some basic intuition as to an answer for that question, neither a historian nor a philosopher can proceed to practice HS or PS, as they would have no idea what it is they are studying! Defining science has tremendous historical value as it answers the question of where to draw the line between history *of science* and general history. Topics like these are critically important and discussing them in &HPS can benefit both the historical and philosophical disciplines. Moreover, these relevant concepts are not limited to demarcation by any means, questions such as: What constitutes scientific change? What are the constituent elements of science? Are there any patterns in science? How is science done? These are all questions that are crucial for both HS and PS. Some of these topics, such as patterns in science, can only be properly studied via both historical and philosophical avenues. There are both diachronic and synchronic studies in &HPS, and these diachronic studies cannot be tackled by PS alone, but require some form of historical research. These mutual philosophical questions will form the basis of my discussion of &HPS.

There is also a more traditional distinction to be made–the normative/descriptive distinction of PS. Within the literature, the normative character of PS has been overwhelmingly stressed and the concept of a descriptive PS has been introduced, but not properly separated from normative PS. PS has had several primary goals regarding science: the interpretation of science; an understanding of science; and most problematically, a prescription for the practice of science. One can do PS by interpreting science, e.g., what does quantum mechanics have to say about determinism? This is of course a topic for &HPS, but it is not a topic that is of particular interest to HS. Understanding science, on the other hand, is quite a different story. HS has a lot to gain from philosophical understandings of science. It can refine what the relevant artifacts are for a study of science, a well-defined taxonomy can be offered for HS, the question of what a scientific narrative consists of begins to have an answer and many more benefits become apparent when PS offers an understanding of science.

Notice, however, that all the benefits HS can draw from a philosophical understanding of science are descriptive, and *not* normative. It can even be said that HS requires some form of descriptive PS (a framework or language) to function, but it surely does not require any PS norms saying how science ought to be done. For example, the question of the Aristotelian-Medieval community's rationality in the face of Galileo's observations has at times been mislabeled as an irrational process based on modern scientific standards, which ignored the episodes historical context (Barseghyan & Shaw, 2022). Normative PS (at least its non-descriptive component) offers little to HS.

Normative PS does however have a place. Although normative PS has often been incredibly historically incorrect and met with criticism[2], it can in principle offer something to scientists. A question such as "what theory *should* be chosen considering the evidence?" can be answered philosophically and potentially even have scientific output. Koertge, for instance, paraphrases Marx as saying "the duty of philosophy of science is to improve scientific practice, not describe it" (Koertge, 1976, p. 367). This certainly adds purpose to normative PS as a more analytic guiding hand for science. Normative PS certainly has a place, but it seems that descriptive PS is the one that matters for &HPS.

I shall label the descriptive philosophy of science, as theories of scientific change (TSC), and the normative philosophy of science as methodology (MTD). The TSC/MTD distinction has been lightly covered by other authors historically, but has not seen a more concrete characterization until more recently.[3] I have found that these distinctions have not been sufficiently clear in separating the normative components of PS from the descriptive, while the TSC/MTD distinction is well-defined in this regard. To gain an intuition for this separation, I provide some examples.

I shall present one illustration of a seemingly pure MTD, an illustration of a mixed case that elucidates why the normative/descriptive distinction in PS has been problematic, and finally an illustration of a seemingly pure TSC case.

Bayesianism and the base-rate fallacy offer an example of a pure MTD. Historically, and currently, it is common that many people commit the base rate fallacy (Dicken, 2013, p. 565). Under Bayesian theory, they ascribe incorrect

---

[2] Just think of Popper's falsificationism, Carnap's verificationism, probability calculus, Kuhnian paradigms, and many more historical philosophies of science that have prescribed certain scientific behaviors. All of these have often been met with distaste from historians of science, actual scientists, and other philosophers of science. However, my aim is not to discuss this topic here.

[3] A recent characterization for the descriptive/normative PS distinction can be found in Scholl (2018). Barseghyan (2015, pp. xi-xvi) discusses the distinction between TSC and MTD. Burian (1977) makes an evaluative/prescriptive distinction, while Brown (1988) makes a normative/naturalized epistemology distinction. Other authors like Laudan (1990) and Giere (1985) are explicit in their normative or descriptive commitments in naturalized epistemology.

probability assessments to situations. Does it matter to Bayesianism that people have commonly committed the base-rate fallacy? No. Bayesianism does not attempt to explain *how* people reason, it is prescriptive in the sense that it tells you what is more probable and perhaps carries the normative baggage that you *should* believe what is more probable. Clearly, Bayesianism is an MTD in this regard, it does not seek to describe science or human reasoning, but it seeks to prescribe arriving at certain conclusions over others based on probabilities (a method). Bayesianism in this sense is not descriptive and hence not a TSC, but a pure MTD.

Next, consider Lakatos's research programmes. Let's focus in on a certain claim: that ad hoc hypotheses (any of the types) are regressive and scientists should avoid them. This is clearly a normative statement which qualifies Lakatos as having created an MTD. But, is that all that Lakatos is saying? No, he is not only saying that scientists should avoid ad hoc hypotheses, but that scientists *do* avoid them. This is a descriptive statement about how science functions. Such a statement could be put into a TSC. Although analyzing Lakatos is merely an illustration in this paper, I invite the reader to consider the whole *Methodology of Scientific Research Programmes* and ask whether the statements are both prescribing norms for science and describing actual science, or simply one of the two. It seems that almost all of Lakatos's claims are both descriptive and prescriptive for science. Lakatos's MSRP thus serves a dual purpose as a TSC and an MTD. This, to me, is why there has been a major conflation of normative/descriptive PS in the HPS literature. This mixing of descriptive and normative PS has been seen in other authors as well such as Popper, Kuhn, and other traditional names in HPS.

On the other hand, Scientonomy from Barseghyan (2015) is a pure TSC. Scientonomy's aim is to be a "science of science" and only offer descriptive claims on what science is, what are its elements, what constitutes scientific change and what the patterns are in scientific change. It does not attempt to prescribe anything to scientists. It follows from a tradition of naturalized philosophy; however, it does not have any normative output. Scientonomy only aims to describe the scientific process, and not prescribe how science ought to be practiced. This makes it a promising venture for those seeking to do &HPS.

3. Food for Thought

In this brief section, I outline an obvious way that HS can be relevant to PS, and that is simply through giving philosophers cases to think about. Garber gives an accurate and charming description of this relevance (Garber, 1986, p. 111):

> It is fair to say that much philosophy of science in the positivist tradition has been armchair philosophy of science; the idea is that we sit down in

a comfortable armchair and think very hard about confirmation, explanation and the like. History of science brings the armchair to the dining table, as it were, and provides food for thought. Despite the fact that it is theory-bound, the history of science can be very valuable in eliciting our intuitions about good and bad science, and in showing us scientific procedures worthy of consideration and practical test that we might otherwise miss.

This kind of interaction between HS and PS works both for a TSC and an MTD. In general, HS provides a "broad range of examples, conceptual tensions and puzzles, which shed new light on present understandings of epistemic terms and might lead to conceptual discoveries" (Schickore, 2002, pp. 453-454). These examples and puzzles may be used in an MTD to find problems or counterexamples of certain accounts, say, for example, of error which lead to a deeper understanding of those terms and how they should be understood and prescribed.

Popular philosophical theses perhaps have stemmed from simply using HS as food for thought. Theses such as the Duhem-Quine thesis, auxiliary hypotheses, incommensurability, the non-cumulative character of science, scientific revolutions, etc. All seem to have stemmed from some inspiration (not necessarily an inductive or hypothetico-deductive inspiration) from the HS.

I hope it is clear that HS can at least serve as food for thought for PS, but it would be a shame if that was all there is to it, and most likely not enough rationale to make an &HPS department or field if this were the only thing HS had to offer for PS. In the next sections, I will cover several approaches on &HPS.

4. Top-down &HPS

Having offered several distinctions that serve as guidelines for what PS is suitable for &HPS, I now take a look at ways history can inform PS by starting with top-down philosophy. The MTD/TSC distinction shall be used extensively to contrast the traditional normative way of practicing philosophy of science, from the descriptive style. Lakatos (1970) introduced a way of integrating history and philosophy of science – his rational reconstructions. Originally inspired by practices in science of taking a hypothesis and making it confront the data, Lakatos perhaps wanted to apply the hypothetico-deductive method of science to the history of science. An aspect of PS for Lakatos was its historiographical input. PS would be able to tell historians what details to focus on in historical episodes and what factors play a role in science and its change. His method for &HPS would go as follows: a methodology (PS) is confronted with the historical data (HS), then it would be possible to judge the PS via the historical reconstruction it produces and how well it fits with the historical data

(Lakatos, 1970, p. 109). Then, *somehow*, the better reconstruction will be chosen as king.

Within this section, I wish to cover a more general version of the so-called rational reconstructions, viz., top-down philosophy. Top-down philosophy is the process where one takes a philosophical thesis and "tests" it against the historical record. I will show how applying a top-down approach to MTDs leads to many problems, while it is far less problematic for TSCs.

For MTDs, one takes a normative prescription and looks at the HS for some form of confirmation/falsification of it. Some scale can be used to compare competing normative theories. This top-down approach has many complications for an MTD. From the previous discussion, one aspect of MTD is that it aims to improve scientific practice. If that is the case, we then arrive at the problem that the HS is clearly in the past, and does not have within itself a novel improvement to scientific practice. At best, one can argue for the revival of a historical scientific practice that has been discontinued. Cases like this however seem to be few and far in between. It seems counterproductive for an MTD to use the HS in order to justify its claims. An MTD wishes to guide science in novel fashions, and not describe previous failures or successes. Thus, it seems that a top-down philosophy cannot be used effectively to test MTDs for their guiding benefits to science, but this is but a small problem compared to the next for an MTD.

The next problem is circularity (Brown, 1988, p. 54; Giere, 1985, p. 333). If one uses some MTD to make a historical reconstruction, and then uses that reconstruction to justify itself, isn't that circular? Is it not the case that the historical reconstruction will be inevitably shoehorned and by this vice, enable its own confirmation? After all, the reconstruction may simply relegate the disconfirming evidence to irrational factors in the historical data.[4] Even if we grant a means to avoiding this problem of shoehorning, what about comparing competing MTDs? What method would we use to choose between the competing MTDs? MTDs *are* our means of choosing between theories, and if we decide to use an MTD in order to pick between MTDs, then our results will always be skewed. We cannot justify the use of an MTD in choosing MTDs as that would be circular. This circularity seems inevitable, but maybe we can make a concession to make it work.

Let us try using a third-party method to pick between MTDs and their respective rational reconstructions. Perhaps we can pick something like the hypothetico-deductive method (à la VPI Project) and use that? If we do this,

---

[4] This is comically seen in Lakatos (1970, p. 107) who relegates the disconfirming history to the footnotes.

maybe we can decide on which reconstruction is better, but the problem remains when we are testing this MTD of theory choice itself. Giere summarizes this issue quite well: "history of science cannot be regarded as providing empirical evidence for a philosophical account of empirical validation" (Giere, 1973, p. 294) because that would presuppose an account of validation. There seems to be no escape from circularity or contradiction within such reasoning, even by using a third-party MTD to arbitrate choice of competing MTDs.

Perhaps the most common objection of testing an MTD with historical reconstructions is the "is-ought" problem (Brown, 1988, p. 55). In the case of a reconstruction, are we not using descriptive historical data to justify/falsify a normative MTD? We are. It is generally taken that no descriptive statements can have bearing on normative statements directly, but I shall leave a more in-depth discussion of this problem for the next section where I tackle bottom-up philosophy.

I now switch my attention to using top-down philosophy for a TSC. Since a TSC is descriptive, it may attempt to mirror some hypothetico-deductive reasoning as seen in science. We take a descriptive thesis, reconstruct some historical episodes using this descriptive thesis and compare it to other reconstructions stemming from different theses. It is clear that as a descriptive science, a TSC will be able to benefit from such a historical study of science, it does not need to guide science like an MTD.

Do we run into circularity when doing top-down philosophy with a TSC? Let us consider a hypothetical scenario. Suppose we have several theses on how theory choice takes place in science and we wish to compare them. What will we use to choose between these competing theses? Evidently, it is some normative criteria. These criteria will be whatever the academics consider to be the method to use at the time, the *modern* method. Notice that this modern method is chosen by the academic community as being the best available at the time. Whatever methods are being tested in history, are the historical methods, which may or may not coincide with the modern method. It is crucial to note that the modern method is not using HS to justify itself, it is already taken for granted, and is merely being used as a tool for generating descriptions of science.

There are, however two remaining problems for TSCs in top-down philosophy: *shoehorning* and *cherry-picking*. Perhaps a historical episode is shoehorned by the TSC to promote itself, or perhaps certain historical episodes are cherry-picked to support a thesis in a TSC. After all, history is vast and one can find a confirming or a disconfirming instance for almost any descriptive statement (Nickles, 1986, p. 256, shares this sentiment). However, one is not limited to a single framework of historical constructions as one can have multiple competing TSCs all reconstructing the same historical episodes, and we can pick between them without a problem. This essentially allows us to deem certain TSCs as better. Thus, shoehorning can be avoided by a proliferation of competing TSCs. Cherry-picking is a problem to all empirical studies. Eventually, methods are developed in order to diminish its effects. Consider something like large historical studies for a TSC, or even randomized ones. Although they are not currently possible, one of the goals of developing a TSC is to be able to get these large coherent databases in order to do these kinds of historical and randomized studies (Scholl & Räz, 2016, p. 230). Both shoehorning and cherry-picking are resolvable over time, as the field of &HPS grows. Although currently these problems are real, the key point here is that one day this cross-disciplinary study of &HPS may become more empirically sound.

Several strides have been taken in this direction. We already have a good example of what competing historical reconstructions could look like. Martinez-Ordaz & Estrada-Gonzales (2018) have already done something of the sort. In their work, they focused on a historical episode in parasitology on the dilemma between biogenesis and heterogenesis of parasitic worms. They reconstruct the same historical episode in three ways called stories A, B, and C. Story A reconstructs this episode as a supporting case of social factors being the only determinants of theory choice. They label this a poor historical construction as it offers little reinforcement for the thesis of social factors being the only factors in theory choice, as it merely says that in the absence of rational factors, social factors were used to choose between biogenesis and heterogenesis. Story B takes a different route and paints a picture of inconsistency toleration between biogenesis and heterogenesis, them being both accepted, but in different empirical domains. They discuss that this is a confirming case, but there is no explicit mention within the historical sources that there was any sort of inconsistency toleration, merely these two (a priori) inconsistent theories seemed to have been both accepted at one point in the medical community. Lastly, in story C, the episode is reconstructed as a case of increase in problem-solving ability. The authors claim that this is the best story as the conclusion is well-reinforced.

As can be seen, from this type of work, we have several competing reconstructions that can all be compared and judged by an &HPS community. Which one is better is not inherent, but to be decided within the community. At the same time, it is not necessary to choose a better reconstruction; one can even keep several so long as they are compatible. So, it seems that MTDs fundamentally struggle with a top-down approach, while TSCs can overcome their deficiencies with time.

5. Bottom-up &HPS

After covering the top-down approach to &HPS, let's have a look at its counterpart. Bottom-up &HPS is the idea to start with historical data and then generalize to some PS. This general concept was often called naturalized philosophy (for example, Giere, 1985; see also Laudan, 1990, who offers a form of naturalized epistemology with normative conclusions). In &HPS, this process involves inferring general statements from historical case studies in science. I shall examine the problems of bottom-up &HPS and notice a fatal flaw of this approach, namely its inability to deal with theory-ladenness on its own.

To start, I will deal with an issue that is common to any conception of bottom-up &HPS, that is, do we only generalize from HS or from what else? The question was brought up whether we should have the "bottom" be composed of purely HS, or if it should include cognitive science, psychology, sociology of science or any other disciplines (Giere, 2011, p. 62). What these disciplines share in common is that they all have some kind of import into the functioning of science. Since we study science as done by humans, cognitive science and psychology may reveal certain behaviors or reasoning patterns that we all use, these patterns could then have some kind of import into how science functions. An example could be our tendency to visualize things and how theories with "better" visual representations came to be accepted by communities. Similarly, sociology might offer many other generalizations relevant to science, e.g., the fact that we publish our research in books or write things down may influence the preservation of scientific ideas. So, to restate the question asked by Schickore (2011a, p. 465), is HS the correct "bottom" to naturalize "up" from? Well, that's up to the &HPS community to decide. However, one can say that there is no reason to exclude any of the given disciplines and they can all complement one another. We need not only do historical generalizations, or only sociological generalizations, all of these generalizations can offer different relevant bases and will (if anything) ameliorate our understanding of science (and its philosophy). So, the choice of "bottom" is not so problematic, but bottom-up approaches have numerous other problems.

I now turn to bottom-up &HPS with MTD. The idea is as follow, we start with some historical data and we want to generalize into some normative statements. The major proponent of such an approach was Laudan (1990), who made a scheme of "hypothetical imperatives" in order to move from historical data to normative statements. The main logic of how this would work is the ability to make theses on "why science worked" (Laudan, 1989, p. 216). If you can say why science worked in one episode, maybe you will be able to say what should be done if you want science to work in the same way again (what means should you employ to reach what end). This is heavily discussed within the literature and I will not focus on Laudan himself. I consider the more general case of simply starting with some historical data and trying to reach a normative output.

I begin with the fundamental problem, the is-ought problem. Historical data is necessarily descriptive, so wouldn't going from historical data to MTD be committing the naturalistic fallacy? The is-ought problem can be expressed as "there is no consistent set of purely descriptive premises D from which a purely normative conclusion N follows, which is not logically true" (Schurz, 1997, p. 11). Bottom-up &HPS seems to be clearly disregarding the is-ought problem which constitutes committing the naturalistic fallacy. But, what if we had some type of bridge principle to go from descriptive propositions to normative ones? One must first appreciate that analytically true bridge principles are not possible (Schurz, 1997, pp. 276-277). What we are left with is synthetic bridge principles, i.e., not necessarily true normative theories that in conjunction with descriptive statements can generate further normative statements (Schurz, 1997, pp. 31, 277-285). These synthetic bridge principles cannot be justified a priori, but we can still accept them and make use of them (just like a scientific theory). Here is a simple example of a synthetic bridge principle: if you promise to do something, then you should do it. If one made use of such a bridge principle in HS, one could find instances when scientists promised something, then conclude that they should do that thing they promised. However, of course, such a bridge principle would be of no interest to most philosophers. Laudan had his own idea of a bridge principle. If one could identify why a scientific mean worked at achieving some scientific end (let's assume that this distinction is unproblematic), then one would be able to get a statement of the form: if you want such and such aim, do such and such mean (a hypothetical imperative). But, why would we choose Laudan's bridge principle over any other one? The decision cannot be a priori, and it would be a shame to call it arbitrary as well. We could of course, under the method of the &HPS community come to agree on a choice of synthetic bridge principle and make use of it. It is conceivable, it would circumvent confronting is-ought problem, but is there any other problem it introduces?

Perhaps there is. From the previous discussion, I noted that one task of MTD was to improve science in some way, but would doing bottom-up &HPS with a synthetic principle be able to achieve that? We run into the same problem as with top-down, an improvement to current scientific practices is almost certainly not going to be found within HS and we will be unable to inductively generalize from historical data in reaching such an improvement. At best, we will be able to generalize some MTDs of the past, find some common trans-historical ones and make historical theses, rather than ones that have any kind of applicable normative output. These trans-historical MTDs come in the form of seeming platitudes, such as predictive power should increase over time. This,

I believe, makes the bottom-up conception a bit dry for an MTD even with a synthetic bridge principle.

The main problem I address is that of the theory-ladenness of historical data. Just like how in scientific observation, we need concepts to make sense of percepts, in HS it is required to have some theory to make sense of the historical record (the documents, artifacts, etc.). As Garber put it, "the history of science (as opposed to our normative judgments about cases in the history of science) cannot be construed as evidence [for any PS]" (Garber, 1986, p. 106). It is appreciated that every historian must work in some framework in order to construct any type of narrative or historical piece, they must assume some kind of *Weltphilosophie* and arrive at their own judgments (Hanson, 1962, p. 574). There is no stand-alone historical data, it must be interpreted via some theory. How can we then proceed to doing HS in a coherent fashion? We must adopt some descriptive theory to describe the HS, namely a TSC. There is of course no a priori TSC to use for neutral historical data. We are forced to be explicit in our commitments, but they may change eventually, and history will then need to be revised or rewritten. Bottom-up &HPS will need to assume at least some form of framework (at least a taxonomy) in order to do any HS. Doing so will no longer be bottom-up &HPS, but instead will involve top-down &HPS since we are assuming some TSC before getting to any HS.

This exactly mirrors the problem in actual science. Science never inducts from neutral data, it's always theory-laden. One needs to start with some conceptual framework to make sense of the data. In the same way, theory-ladenness in &HPS needs to start with some conceptual framework. This of course does not mean we are trapped in this framework, nor that it is the only one we can use. As before, we can reconstruct things in different ways and accept several reconstructions. One important thing to note is that historians frequently omit being explicit with their given framework (Hull, 1992, p. 472; historians typically assume some theory of rationality in passing and use it to explain changes in intellectual history). Having a TSC is indispensable in this regard, as having a consistent vocabulary, precise meanings to words used within a historical narrative, an understanding of the epistemic processes present in science, etc. are all crucial components that are lacking in particularist approaches to HS. In order to properly make use of bottom-up &HPS, we will need to couple it with a top-down approach.

6. Iterations

Both the top-down and bottom-up methods of &HPS seem to have been based on some form of hypothetical-deductivism and inductivism respectively from the sciences. But, are these methods truly how science proceeds? Schickore would say that top-down and bottom-up are merely modes of presentation of scientific research rather than a mode of the production of scientific knowledge (Schickore, 2011a, p. 473). It is instead proposed that science actually follows something along the lines of an iterative fine-tuning of theoretical concepts that interpret observational data, which then in turn gives insight to new theoretical concepts and the process repeats itself. I shall call such an iterative production of knowledge simply as "iterations".[5] I shall present several examples of such iterations, including a scientonomic conception.

Let us start with Mill's inductivism as presented in Knowles (2003). Here Knowles applies Mill's inductivism in the context of scientific norms to show that they are fine-tuned in the same fashion as observations and hypotheses are fine-tuned (Knowles, 2003, pp. 71-73). That is, one has a working hypothesis, produces some observations with that hypothesis, evaluates those observations and uses them to justify the next hypothesis. Similarly, this can be applied to norms on the next (meta-)level up. One has some norms and evaluates the success of the hypotheses accepted under such norms, and the results of this evaluation justify new norms. Such a process is an iteration between norms and successes of hypotheses. In Figure 2.1, directions of justification/support are depicted by the sloping/vertical arrows, while the bold horizontal arrow indicates "the progress (assumed to be) being made over time" (Knowles, 2003, p. 71):

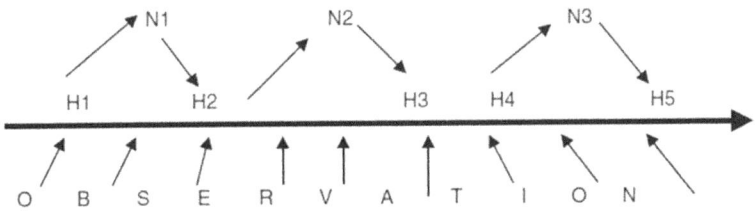

Figure 2.1: Knowles' diagram

---

[5] This steals from Hasok Chang's naming, but I believe it is a nice picture of how it works. Similar conceptions of iterations have been studied by numerous authors, and they all have a similar structure.

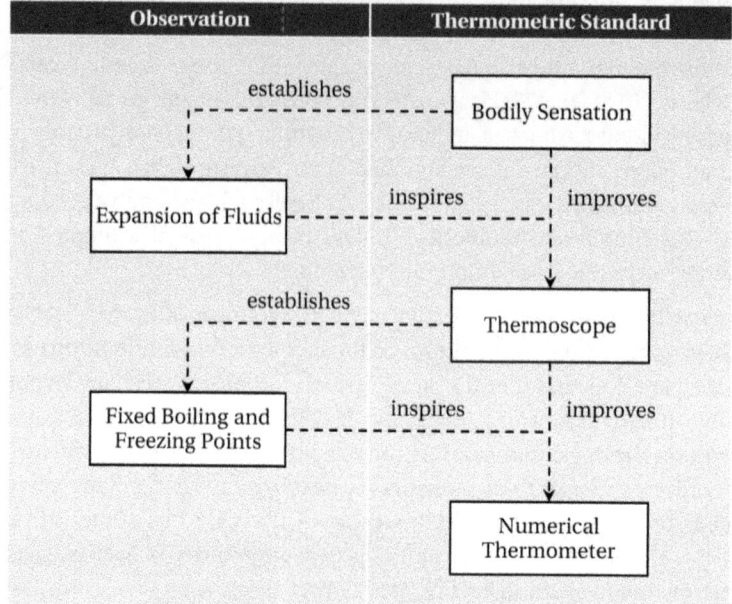

Figure 2.2: A general form of iteration

Chang (2011) also presents us with an iteration. He calls it an epistemic iteration and applies it to the nomic problem of measurement. In the context of temperature measurement, the epistemic iteration progresses as follows. One starts with a method to measure temperature, then tests it out (perhaps in comparison to other methods) and evaluates the results, which inspire fine-tunings of the next measurement method, and the process repeats. The idea is that over each iteration, you fine-tune your temperature measurement method even further, eventually arriving at more and more suitable measurement methods. Epistemic iterations are not limited to measurement methods, one can apply them to any process of inquiry. You start with some ill-defined methods or concepts, test them out, re-evaluate your methods or concepts based on the results of the previous testing, and repeat. Hence, Chang has a general form of iteration (Figure 2.2).

Scholl & Räz (2016) also have their own version of helical iterations called cyclical HPS. In this case, abstract philosophical concepts enlighten historical episodes which provide concrete results that inform further abstract concepts, and the process repeats. Although they do repeat Chang's idea, they now show how it can be applied to &HPS in specific.

Next is Schickore's hermeneutic-critical style. It can be summarized as follows (Schickore, 2011a, p. 471):

Initial case judgments – judgments that identify portions of the historical record as noteworthy – and provisional analytic concepts are gradually reconciled until they are brought into equilibrium.

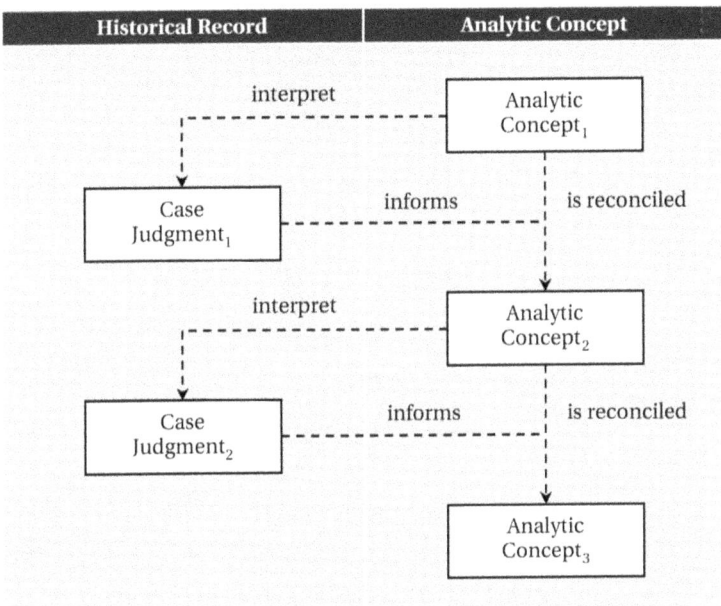

Figure 2.3: Iteration in &HPS

Although Schickore dismisses top-down and bottom-up philosophy as modes of presentation, I believe there is a clear connection between those two and the alternative iterative process Schickore suggests. It seems that by taking a provisional concept, one is effectively doing some form of top-down &HPS, at least locally, as they assume a framework and use it to interpret an episode. This interpretation is then inductively used to refine the concepts, this step consists of the bottom-up portion of Schickore's iterative process. As such, Schickore's hermeneutic-critical style (Figure 2.3) seems to be a sequence of local top-down and bottom-up steps, used to gradually refine philosophical concepts using historical research.

Lastly, I present the scientonomic conception of iterations. The second and third scientonomic laws predict an iterative process. By the second law, theories come to be accepted through a satisfactory assessment by the method of the time (Patton, Overgaard, & Barseghyan, 2017). Then, by the third law, methods come to be revised through the newly accepted theories (Sebastien, 2016). This illustrates an iterative process where methods are fine-tuned by newly accepted theories. An example from clinical epidemiology is visualized in Figure 2.4.

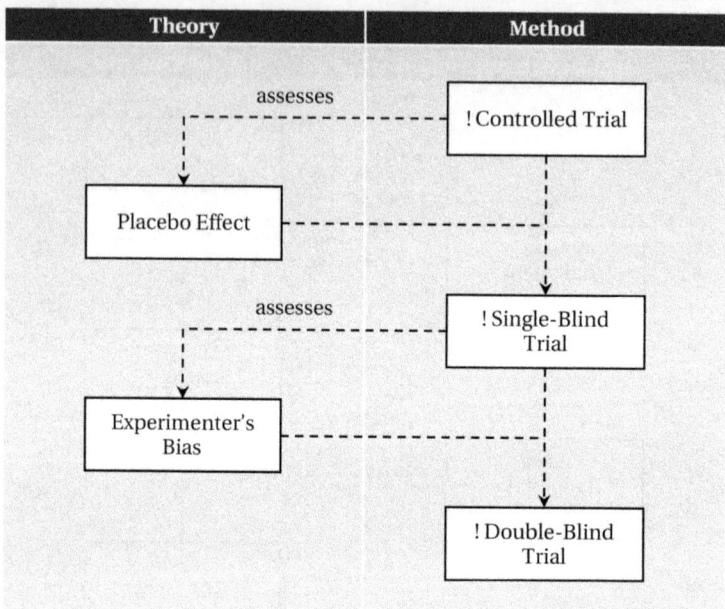

Figure 2.4: An example of iteration from clinical epidemiology

We see here that the method of control trials is able to identify there being a placebo effect. This in turn inspires a new method, namely that of a single-blind trial, making for the first iterative development. The single-blind trial is then able to identify the experimenter's bias, which makes the double-blind trial deducible by the third law. This makes for two iterative steps of scientonomic iterations.

Overall, I propose a general conception of iterations inspired by these examples. All the examples mimic a form of scientific production of knowledge, as an iterative interaction between theory and evidence. A theory is used to interpret evidence, which is then used to (dis)confirm the theory itself, or other competing theories. Iterations are summarized diagrammatically in Figure 2.5.

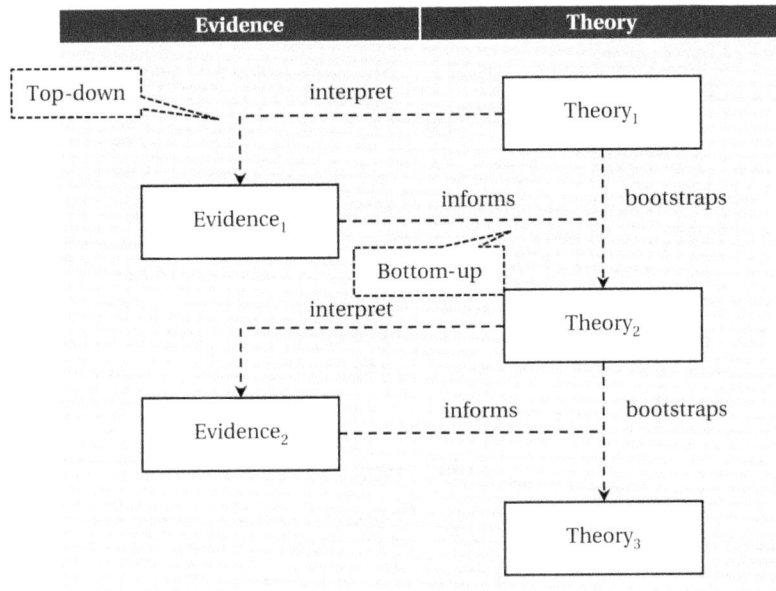

Figure 2.5: Iterations summarized

Notice that we have two types of steps here. Theories may interpret evidence; this constitutes a top-down step of the iteration. Then, this evidence may be used to inform new theories; which constitutes a bottom-up step of the iteration. This makes the iterative production of knowledge not restricted to either top-down or bottom-up approaches, but actually integrates the two in a seamless fashion. We avoid the problem of theory-ladenness of bottom-up approaches, as the prior top-down step sets up a framework to interpret the evidence. We further are able to appropriately generalize from the evidence without having to propose new hypotheses seemingly out of nowhere, the hypotheses are informed by prior evidence. Lastly, we see how prior theories may build upon themselves in what has been typically called a 'bootstrapping' step, though it should be clear that this is a misnomer in the case of iterations as interpreted evidence (along with theories) plays a role in this step. This process generalizes the iterative production of knowledge.

All of these approaches to iterations have been able to fruitfully study either science itself, or offer means of studying science. Iterations avoid the nefarious problems present for top-down and bottom-up approaches to studying the history of science, and have been shown to offer interesting takes on historical episodes. I hope that this recognition of the iterative processes present in studies of science is able to further progress &HPS as a field. I believe that through iterations between PS, acting as theory, and HS, acting as evidence,

&HPS can be united into fruitful areas of study, as the exemplary authors discussed have proven.

7. Conclusion

In explicitly characterizing the relevant PS for &HPS I proceeded to study the methods by which the HS informs PS. I discuss HS as food for thought for PS, a top-down approach to &HPS, a bottom-up approach to &HPS, and finally culminate in a discussion of iterations. I hope to have shown that iterations are present both in science and in studies of science. Moreover, I hope to have shown that iterations serve as an interesting instrument for studying epistemic continuities as they identify a powerful mechanism by which the progression of knowledge can be facilitated.

Bibliography

Arabatzis, T. & Schickore, J. (2012). *Introduction: Ways of Integrating History and Philosophy of Science.* Perspectives on Science, 20(4), 395-408.

Barseghyan, H. (2015). *The Laws of Scientific Change.* Springer.

Barseghyan, H. & Shaw, J. (2022). Integrating HPS: What's in it for a Philosopher of Science? In this volume, 41-65.

Brown, H. (1988). Normative Epistemology and Naturalized Epistemology. *Inquiry*, 31, 53-78.

Burian, R. (1977). More than a Marriage of Convenience: On the Inextricability of History and Philosophy of Science. *Philosophy of Science*, 44(1), 1-42.

Chang, H. (2011). Beyond Case-Studies: History as Philosophy. In Mauskopf & Schmaltz (Eds.) (2011), 109-124.

Cohen, R. S., Feyerabend, P., & Wartofsky, M. W. (Eds.) (1976). *Essays in Memory of Imre Lakatos.* D. Reidel.

Dicken, P. (2013). Normativity, the Base-Rate Fallacy, and Some Problems for Retail Realism. *Studies in History and Philosophy of Science*, 44, 563-570.

Garber, D. (1986). Learning from the Past: Reflections on the Role of History in the Philosophy of Science. *Synthese*, 67(1), 91-114.

Giere, R. (1973). History and Philosophy of Science: Intimate Relationship or Marriage of Convenience? *The British Journal for the Philosophy of Science*, 24(3), 282-297.

Giere, R. (1985). Philosophy of Science Naturalized. *Philosophy of Science*, 52(3), 331-356.

Hanson, N. (1962). The Irrelevance of History of Science to Philosophy of Science to Philosophy of Science. *The Journal of Philosophy*, 59(21), 574-586.

Hull, D. (1992). Testing Philosophical Claims about Science. *PSA: Proceedings of the Biennial Meeting of the Philosophy of Science Association*, 1992, 468-475.

Knowles, J. (2003). *Norms, Naturalism and Epistemology.* Palgrave Macmillan.

Koertge, N. (1976). *Rational Reconstructions.* In Cohen, Feyerabend, & Wartofsky (Eds.) (1976), 359-369.

Lakatos, I. (1970). History of Science and Its Rational Reconstructions. *PSA: Proceedings of the Biennial Meeting of the Philosophy of Science Association*, 1970, 91-136.

Laudan, L. (1989). The Rational Weight of the Scientific Past: Forging Fundamental Change in a Conservative Discipline. In Ruse (Ed.) (1989), 209-220.

Laudan, L. (1990). Normative Naturalism. *Philosophy of Science*, 57(1), 44-59.

Martinez-Ordaz, M. & Estrada-González, L. (2018). May the Reinforcement Be with You: On the Reconstruction of Scientific Episodes. *Journal of the Philosophy of History*, 12, 259-283.

Mauskopf, S. & Schamltz, T. (Eds.) (2011). *Integrating History and Philosophy of Science: Problems and Prospects*. Springer.

Nickles, T. (1986). Remarks on the Use of History as Evidence. *Synthese*, 69, 253-266.

Patton, P., Overgaard, N., & Barseghyan, H. (2017). Reformulating the Second Law. *Scientonomy*, 1, 29-39.

Richardson, A. (1992). Philosophy of Science and Its Rational Reconstructions: Remarks on the VPI Program for Testing Philosophies of Science. *PSA: Proceedings of the Biennial Meeting of the Philosophy of Science Association*, 1992, 36-46.

Sebastien, Z. (2016). The Status of Normative Propositions in the Theory of Scientific Change. *Scientonomy*, 1, 1-9.

Schickore, J. (2002). (Ab)Using the Past for Present Purposes: Exposing Contextual and Trans-Contextual Features of Error. *Perspectives on Science*, 10(4), 433-456.

Schickore, J. (2011a). More Thoughts on HPS: Another 20 Years Later. *Perspectives on Science*, 19(4), 453-481.

Schickore, J. (2011b). What Does History Matter to Philosophy of Science? The Concept of Replication and the Methodology of Experiments. *Journal of the Philosophy of History*, 5, 513-532.

Scholl, R. & Räz, T. (2016). Towards a Methodology for Integrated History and Philosophy of Science. In Scholl & Sauer (Eds.) (2016), 69-91.

Scholl, R. & Sauer, T. (Eds.) (2016). *The Philosophy of Historical Case Studies*. Springer.

Scholl, R. (2018). Scenes from a Marriage: On the Confrontation Model of History and Philosophy of Science. *Journal of the Philosophy of History*, 12, 212-238.

Schurz, G. (1997). *The Is-Ought Problem*. Springer.

Chapter 3

Integrating HPS:
What's in it for a Philosopher of Science?

Hakob Barseghyan

University of Toronto

Jamie Shaw

University of Toronto

Abstract: Many have struggled to identify the proper way(s) that normative philosophical claims about science can benefit from history. The primary worry here has been that deriving philosophical 'oughts' from historical facts would commit the naturalistic fallacy (Schickore, 2011). The task of this paper is to introduce a novel solution to this problem. Specifically, we claim that the emerging field of scientonomy provides a promising avenue for how philosophy of science may benefit from the history of science. By taking descriptive findings and coupling them with additional normative premises, philosophers of science can draw normative methodological conclusions which can guide future scientific practices. Moreover, it is sometimes thought that philosophical claims about science are invariably local due to the diversity of scientific practices. While acknowledging this disunity, we show how a general theory of scientific change is possible and how it can be used to inform the normative philosophy of science. Thus, we aim to outline a viable path for integrated history and philosophy of science that does not relinquish normativity and avoids the problem of cherry-picking which has plagued general accounts of science (Chang, 2011; Mizrahi, 2015).

Keywords: integrated history and philosophy of science; normative; descriptive; philosophy of science; history of science; naturalistic fallacy

***

1. Introduction

Since the historical turn, there has been a great deal of anxiety surrounding the relationship between the history of science and the philosophy of science. Surprisingly, despite six decades of scholarship on this topic, we are no closer to achieving a consensus on how these two fields may be integrated. However, recent work has begun to identify the crucial issues facing a full-fledged account of integrated HPS (Domski & Dickson (Eds.), 2010; Schickore, 2011; Mauskopf & Schamltz (Eds.), 2011; Herring et al. (Eds), 2019). We contend that the inability to deliver on a model of integrated HPS is partly due to an insufficient appreciation of the distinction between normative and descriptive accounts of science, an over-emphasis on individual case studies, and the lack of a general theory of science to mediate between historical data and philosophical conceptions of science.

In this paper, we provide a novel solution to this conundrum. We claim that the emerging field of *scientonomy* provides a promising avenue for how the philosophy of science may benefit from the history of science. We begin by showing that much of contemporary philosophy of science is ambiguous as to whether it attempts to answer normative or descriptive questions. This ambiguity has led to many attempts to cherry-pick case studies and hastily draw normative methodological conclusions. Against this, we claim that descriptive and normative questions should be clearly separated so that we may show how descriptive history of science may benefit normative philosophy of science. Specifically, we show that a general theory of scientific change is required to mediate between individual findings of the history of science and normative considerations of the philosophy of science. Such a general descriptive theory is necessary to avoid the problem of cherry-picking and the temptation to shoehorn historical episodes into the normative confines of a certain methodology. The main aim of scientonomy is to understand the mechanism that underlies the process of changes in both theories and the methods of their evaluation. We demonstrate how a gradual advancement of this general descriptive theory can have substantial input for the normative philosophy of science and turn scientonomy into a link between the descriptive history of science and the normative philosophy of science.

2. What is It that Philosophers of Science Do?

To know how the history of science may benefit the philosophy of science, we must first know what philosophers of science do. At first glance, this topic seems excessively broad. Philosophers engage in many kinds of projects and, given the growing interdisciplinary interactions of philosophy with other fields of inquiry, the borderline between "philosophy" and "non-philosophy" is becoming increasingly difficult to discern. Still, there are some common threads that bound many philosophers of science together.

One such thread is *metaphysical* questions. What sort of processes, entities, and/or relations make up the world at a fundamental level? What is the nature of space and time? Are there laws of nature or just local patchwork of facts? Are there individual objects or only structures? Are social groups, organisms, ecosystems, or consciousness genuinely emergent entities? These metaphysical questions are of great importance specifically for deciphering the ontological commitments of particular scientific theories. Obviously, this task will require drawing on the findings of the history of science. However, there is another family of questions that philosophers are concerned with when consulting the history of science. Historically oriented philosophers have often been interested in a variety of *epistemological* and *methodological* questions. When, if ever, are we justified in believing in unobservable entities? Should we interpret our theories as elliptical descriptions of observational propositions or as literal depictions of the world? What counts as a proper explanation? Under which conditions are we justified in accepting a theory? While it is common for historically oriented philosophers to back up their positions by referring to the history of science, the exact relation between the history and the philosophy of science remains unclear.

Part of the issue has to do with the ambiguity of the questions that historically inclined philosophers of science try to address: what do they hope to achieve as a result of their investigations? Namely, it is unclear as to whether their findings are meant to be *descriptive* or *normative*.[1] This ambiguity is, of course not new. Consider, for instance, Feyerabend's criticism of Kuhn (Feyerabend, 1970, p. 132):

> Whenever I read Kuhn, I am troubled by the following question: are we here presented with *methodological prescriptions* which tell scientists how to proceed; or are we given a *description*, void of any evaluative element, of those activities which are generally called 'scientific'? Kuhn's writings, it seems to me, do not lead to a straightforward answer. They are ambiguous in the sense that they are compatible with, and lend support to, both interpretations.

---

[1] *Normative* propositions involve value judgements and say how something *ought* to be, what's good/bad, right/wrong, imperative/permissible/forbidden, etc. In contrast, *descriptive* propositions attempt to describe and/or explain how and why things *are* now, *were* in the past, or predict how and why they *will* be in the future (Barseghyan, 2015, pp. 11-21).

At times, it seems as if Kuhn is merely trying to *describe*, in a philosophical kind of way, what scientists do (e.g., what practices they engage in, how they organize their communities) and what they think about what they do. At other times, it seems as if Kuhn is *appraising* particular practices: scientists *ought* to proceed within normal science until a revolution is forced upon them by a growing crisis, it is *rational* for scientists to shelve anomalies, etc. Kuhn's response to Feyerabend is unfortunately unsatisfying (Kuhn, 1970, p. 237; see also Kuhn, 1977, pp. 3-21):

> [It] should be read in both ways at once. If I have a theory of how and why science works, it must necessarily have implications for the way in which scientists should behave if their enterprise is to flourish. The structure of my argument is simple and, I think, unexceptionable: scientists behave in the following ways; ... in the absence of an alternate mode that would serve similar functions, scientists should behave essentially as they do if their concern is to improve scientific knowledge.

This reinforces the ambiguity Kuhn is being charged with: he jumps from what scientists happen to do to what they should do. In order to make his argument sound, Kuhn had to tell us what's so good in the way science is actually done, which would require a normative analysis.[2] This ambiguity persists to this day. Take the following passage as an example (Mitchell, 2009, p. 2):

> As I will discuss throughout this work, there is a mismatch between the current sciences of complex behaviors and standard philosophical criteria for scientific knowledge. It is my view that new approaches to knowledge and action are required to understand and manage the complexity of nature. These practices can be found in contemporary sciences.

There are many subtle ambiguities which are clouded here, making it difficult to discern the normative and descriptive components. In the first sentence, we hear that there is a "mismatch" between what scientists do and what "standard philosophical criteria" tell them to do. But then how would this be a problem

---

[2] This criticism, however, is not unique to Kuhn. For example, Popperians strived to uncover the normative content of logical positivism, which was not obvious given the way they presented their claims (see Popper, 2002, p. 12; Bar-Hillel, 1968; Lakatos, 1968). An exception to this trend is Reichenbach (1938, ch. 1). Carnap (1966, p. 249) broaches this topic but does not provide a definitive answer. In the contemporary literature, there remain but a few who clearly espouse whether their views are normative or descriptive (e.g. Chang, 2012, p. 269).

for these "philosophical criteria"? If they are *normative*, then they cannot be refuted by the fact that there is a "mismatch" between them and practice. The *practice* should be reformed, not the normative methodology. This makes the argumentative move in the second sentence confusing. Why should we construct a new methodology because some practices deviate from prior norms? Is the purpose of this to establish a more comprehensive *description*? If so, this seems to be a different kind of project than that of prior normative accounts. Or, rather, is the goal to come up with a more suitable *normative* methodology? If so, then the appeal to practices is circular since it *assumes* that those practices are instances of sound reasoning in the first place. This shows how the confusion between the normative and descriptive can obfuscate the nature of an argument.

More generally, the *goals* of many papers in the philosophy of science are ambiguous. We often hear that the goal is a "better understanding" of what a model is, or an "account" of explanation, or a "framework" for understanding inconsistency toleration in science. On the one hand, philosophers can be interpreted as addressing *descriptive* questions. If that's what they are doing, then their descriptive accounts should be assessed by examining to what extent they mirror actual scientific practice. On the other hand, philosophers can also be understood as addressing *normative* questions, i.e., as aiming to provide a *methodology* of what beliefs are warranted, what inferences are legitimate, or what arguments made by scientists are valid. In that case, historical episodes can only provide *examples* or *illustrations* of their normative methodologies. This ambiguity is especially unfortunate since, without resolving it, it is unclear what role history can play in relation to the philosophy of science.

It would be impossible to resolve each instance of this ambiguity and to determine the aims of each of these investigations. Rather, we will proceed by clearly distinguishing between the normative and descriptive questions. Specifically, we will continue with the conviction that philosophy of science has important *normative* functions to fulfill. To be sure, any historically oriented philosopher would agree that we need a *descriptive* account of the process of scientific change to answer *normative* philosophical questions; this, however, is not to be taken as a sign that philosophy of science is itself a descriptive enterprise.[3] Our task for the remainder of this paper is to see how normative philosophy of science may best benefit from descriptive studies of science.

---

[3] To be clear, many philosophers have aimed to understand what tools from philosophy may aid historians and what tools from history may aid philosophers. In other words, they seek to construct a division of labor where practicing historians and philosophers collaborate to

3. How does Philosophy of Science Rely on the History of Science?

When the historical turn introduced the use of history as a vital component of the philosophy of science, its exponents proposed extremely general characterizations of science. Lakatos, Laudan, and others used a plethora of case-studies to uncover the *general* method underlying the process of scientific change. It was one of the central ideas of the historicist tradition that different normative methodologies can be tested against the historical record. While it was never really agreed as to how exactly this testing should be conducted, the common denominator of many historicists was that different methodological dicta ought to be tested against a set of intuitively selected historical transitions that we all consider rational (e.g., from geocentrism to heliocentrism, from Aristotelian to Newtonian natural philosophy, from classical mechanics to general relativity, etc.). A methodology would be acceptable only if its normative verdicts coincided (by and large) with those actually made by scientists (Lakatos, 1971; Lakatos & Zahar, 1976; Wykstra, 1980, p. 212; Laudan, 1986). Thus, according to this *meta*-methodology, it is the task of a methodology to demonstrate that its verdicts square with our basic intuitive value judgements.[4]

One assumption underlying this approach was the idea that the preselected set of transitions is genuinely representative of the general population of transitions. That is, by reconstructing the methodological rules that drove the transitions in the intuitively preselected cases, we can make a legitimate claim about the broader applicability of those methodological rules to *all* rational historical episodes. This assumption is based on a more fundamental assumption – the idea that there *is* one fixed set of methodological dicta – "*the* scientific method" – employed during all genuinely scientific transitions. It is obvious nowadays that both of these assumptions are deeply problematic: since we now know that different domains of inquiry often employ different methods and since those methods change through time, any generalization from preselected historical case studies is extremely dubious.

While few philosophers of science would openly accept the meta-methodology of Laudan and his colleagues, the case-study approach is still the meta-methodology of choice for many historically oriented philosophers of science. It is quite common for a contemporary essay to proceed by providing one or two

---

produce interesting research (Chang, 1999; Schickore, 2011). While such research is extremely important from a practical perspective, the question that we are concerned with here is what the relationship between the history and philosophy of science amounts to.

[4] The culmination of this approach was the infamous VPI project, which failed for a number of reasons that are beyond the scope of this paper (Donovan, Laudan, & Laudan (Eds.), 1992; for criticism see Nickles, 1986, 1989; Richardson, 1992).

in-depth case studies and generalizing methodological conclusions from them (e.g., Nickles, 1995; Allchin, 2003; Bueno, 2008). The case-study approach has raised some eyebrows and, specifically, has led to accusations of cherry-picking historical cases to warrant foregone philosophical conclusions (Pitt, 2001). Chang puts this point quite straightforwardly (Chang, 2011, p. 109):

> What can we conclude from a mere handful of case studies?... The field of HPS has witnessed too many hasty philosophical generalizations based on a small number of conveniently chosen case studies. I believe that the neglect to clarify the nature of the history–philosophy relationship in case studies has contributed decisively to a widespread disillusionment with the whole HPS enterprise.

Following up on this worry, Mizrahi writes that "trying to undermine scientific explanations that are based on defeasible inferences by cherry-picking a few cases is bound to fail unless we have good reasons to believe that the cherry-picked cases are representative of the general population" (Mizrahi, 2015, p. 133). This is not merely an accusation of bias against the researchers, but a substantive methodological issue for HPS. To know that a given selection of case studies can reveal something *general* about the methodological dicta guiding the process of scientific change, we need to assume that these cases are genuinely representative of the general population. But that requires some general knowledge of the process of scientific change to begin with. Thus, this approach either assumes some a priori general knowledge concerning the process of scientific change or is blatantly circular.

As a rule, we cannot generalize from case studies and conclude, from those generalizations, that those case studies are exemplary of the general population. Rather, we need a much more comprehensive knowledge of the process of scientific change to *ensure* that we are not cherry-picking. To be sure, many philosophers are likely honest and do not intentionally cherry-pick their examples. However, methodologically speaking, we cannot be sure that this isn't the case without a general knowledge of the process of scientific change.

In addition, this approach makes a deeply problematic assumption that there is a core set of methodological rules employed in all domains of legitimate science at all times. Say we conducted a meticulous study of a certain historical transition which revealed that the scientists of the time did indeed expect new theories to have confirmed novel predictions to be accepted. If we could assume that the same set of requirements is employed during *all* cases of theory evaluation throughout history (for example, due to some inherent rationality of human agents), then we could, in principle, generalize our finding and make a legitimate case that the methodological rule "a theory is acceptable only if it has confirmed novel predictions" is part of the fixed and universal method of science. Needless to say, such an assumption doesn't hold water: there is abundant evidence that methodological requirements of scientists have changed through time (Kuhn, 1970; Feyerabend, 1975; Shapere, 1980; Laudan, 1984). This goes not only for changes in methods of specific domains, such as the transition from single-blind to double-blind method in clinical epidemiology, but also – importantly – for changes in most fundamental methodological dicta, such as the transition from the early modern method of intuition schooled by experience to the method of hypotheses circa 1700 (Barseghyan, 2015, pp. 132-152). Given that the methods of theory assessment are constantly in flux, we cannot generalize from select case studies to all cases. Additionally, we cannot extrapolate from *past* cases to *future* science. Since normative philosophy of science must inherently be forward-looking, this makes it impossible in principle to use case studies to *directly* support any normative methodological conclusions. As nicely summarized by Dudley Shapere long ago (Shapere, 1977, p. 499):

> It is not our purpose to make generalizations about science on the basis of examination of single, isolated cases, or even, necessarily, on the basis of characteristics possessed in common by a number of cases. This is not merely because generalizations on the basis of a single or few cases are always dangerous, but because generalizations we seek are of a different sort. They have to do with the dynamics of rational change, which make characteristics discernible at any one point of science altered or obsolete at a later stage, and therefore not suitable as bases of generalization.

What's left of the case-study approach if we accept these problems? It appears as though we do not have strong reasons to generalize very far from the initial case study. As Chang puts it, "we are condemned to either unwarranted generalizations from historical cases or entirely 'local' histories with no bearing on an overall understanding of the scientific process" (Chang, 2011, p. 111). Thus, any generalization from case studies concerning normative methodological dicta can be, at best, at a very low level.[5]

It is not surprising, therefore, that many analytic philosophers these days opt to engage in a brand of philosophy of science that is anything but historical (see Chang, 1999, p. 420). They revert to the classical pre-Kuhnian tradition where

---

[5] Some have claimed that we should rest content with these low-level generalizations where we can respect the motivation of abandoning a general depiction of the history of science while having a general *enough* theory to motivate context-specific norms (Burian, 2001). This approach, however, faces its own difficulties (see Schickore, 2011, p. 469).

the history of science had virtually no input in the assessment of normative methodologies. The key assumption guiding this approach is the idea that scientists may be mistaken, even about the nature of their own trade; in fact, "the whole of science may err" (Popper, 2002, p. 5). As such, philosophical reasoning must correct these mistakes when and where they exist. This is a purely anti-historicist approach where philosophers of science devise normative methodologies established by philosophical reasoning and intuition. On this view, the historical case studies can only be elucidations of general methodological principles and cannot be used to support or combat them. This can lead to the unsatisfactory conclusion that *philosophies of science* may have absolutely nothing to do with *science itself.* As Feyerabend expressed this point, philosophical positions without any contact with the history of science would be mere "castles in the air" (Feyerabend, 1970, p. 127).[6] Even worse, this would make the choice between competing philosophical positions, insofar as they are coherent, arbitrary.[7]

Thus, we face a dilemma, both forks of which are dead ends. Attempts to base normative methodological conceptions of the philosophy of science on the data of the history of science seem unsatisfactory due to the problem of cherry-picking and a deeper issue with the changeability of scientific methods. However, a refusal to take the historical data into account is equally unsatisfactory, as it can lead to a philosophy of science that has virtually nothing to do with actual scientific practice. Both approaches are untenable. Luckily, these are not the only two options. In what follows, we will outline an approach that allows for philosophers to make substantive use of the history of science while avoiding the problems of these approaches.

4. What do we Suggest?

These days, when a historian of science delves into a historical episode, she primarily aims at reconstructing the complex nuances of that episode to the best of her ability while simultaneously trying to avoid value judgements concerning the actions of the respective historical agents. Essentially, her questions are *descriptive.* What were the beliefs, methods, and practices of the relevant historical agents? How exactly did those beliefs, methods, and practices change during the period under study? What were the relevant factors that contributed to the transitions in those beliefs, methods, and practices

---

[6] It may, of course, be possible that a philosophical position *happens to be* similar to scientific practices. However, such a similarity would be incidental and irrelevant to the soundness of the position itself.

[7] Popper (2002, p. 59) himself recognized that there are other philosophical positions that are as equally philosophically defensible as his own falsificationism (e.g. conventionalism).

during that time period? Our historian is expected to shed light on the intricate tapestry of the past events as they unfolded, notwithstanding any normative interests she may have regarding her subject matter. What matters is the accuracy of her depiction of an episode; whether this depiction can also serve some normative goals and guide future science is irrelevant to the justification of her descriptive account. To be clear, her normative assumptions may affect what part of science she is interested in, but they would not lend any credence to whether or not her description is accurate. In contrast, the methodologist's main aim is *normative*, as she tries to evaluate the methods and practices of contemporary science, and – if necessary – suggest better methods and practices moving forward.

We agree with Lakatos (1971) that the historian needs some theoretical framework to explain historical transitions.[8] Yet, we don't think this framework can be provided by any *normative* methodology, since whatever drove those transitions would be the theories, methods, and practices of the actual historical agents. Importantly, the methods employed by these historical agents may be distinct from the methodological dicta of some contemporary philosopher of science. Given what we know about the changeability of scientific methods, this much is clear. What the historian cannot do without is some *general descriptive* assumptions about the mechanism of scientific change. If the history of science aims to be anything more than a mere chronology of events, it must rely on some general assumptions about the underlying dynamics. Consider, for example, a typical historical passage (Home, 2003, p. 363):

> Newton's experiments with prisms and, more particularly, the conclusions he drew from them caused a flurry of controversy when they were first published in the 1670s. In France his ideas were actually rejected for a generation because his experiments could not be replicated, but in the 1710s they came to be fully accepted.

The tacit assumption here is that *replicability* is somehow important for the acceptance of a theory. What is never clarified is whether this assumption is meant to be a *general* feature of science or merely one of the requirements of the method of *that* community at *that* time. If it is the latter, then there is another tacit assumption that theories remain unaccepted by an epistemic agent unless they meet the requirements of that agent (in this case, the

---

[8] There are other aspects of Lakatos' view that we are suspicious of, including his claim that philosophers may 'fabricate' historical examples. However, we do not have the space to address these concerns here. See Kuukkanen (2017) for further discussion.

requirement of replicability). This assumption, as we have come to learn, is by no means trivial (Patton, Overgaard, & Barseghyan, 2017). Here is another similar passage (Wilson, 2003, p. 329):

> Kepler's so-called second law – the equable sweeping out of area by the Sun-planet vector – … had been generally rejected early because of mathematical difficulties it entailed.

Here, the tacit assumption is that mathematical difficulties can lead to the rejection of a theory. Once again, it is unclear whether the author takes this as a *general* feature of science or merely as one of the ingredients of the method of theory evaluation employed by *that* community at *that* time. If it is the latter, the author assumes that epistemic agents accept new theories only when these theories meet *their* expectations. It is safe to say that any historical narrative is necessarily awash with such *general* assumptions and there is nothing surprising about this (Barseghyan, 2015, pp. 72-79). Unless we make these assumptions explicit, we cannot scrutinize them or even understand what they are. As a result, we won't be able to evaluate the soundness of the respective historical narratives.

Crucially, these general assumptions are attempts to answer a range of *general descriptive* questions about the process of scientific change. What's the mechanism of theory acceptance? What's the underlying dynamics of method employment? How does theory or method rejection take place? In addition to questions concerning the dynamics of changes in theories and methods, there are questions concerning the ontology of the process. What are the basic epistemic *elements* that undergo scientific change, i.e., is it just theories and methods, or is it also questions, reasons, values, etc.? What are the types of epistemic *stances* that an epistemic agent can take towards an epistemic element (e.g., acceptance, use, pursuit, employment, etc.)? Addressing these and other similar questions concerning the ontology and dynamics of scientific change amounts to developing a general *descriptive* theory of scientific change.

Thus, there are three distinct groups of questions which, albeit interconnected, concern different aspects of scientific change:

1. **Normative methodological** questions concerning methods and practices that *ought to* take place in science.

2. **Descriptive historical** questions concerning the peculiarities of theories, methods, and practices in individual historical episodes.

3. **Descriptive theoretical** questions concerning the underlying
    mechanism of changes in theories, methods, and practices.

We believe that the key to our conundrum is the latter group of questions, which
traditionally has been touched upon mostly in between the lines by the reluctant
historian, or has been blended with normative methodological questions by the
eager philosopher (Barseghyan, 2015, pp. 12-20). A system of answers to
descriptive theoretical questions, we contend, is the missing link between the
historical data and normative-cum-philosophical methodologies.

Scientonomy is an emerging attempt to address descriptive theoretical
questions concerning the ontology and dynamics of scientific change in a
systematic, piecemeal, and transparent fashion. As a descriptive enterprise,
scientonomy doesn't presuppose the existence of any fixed or universal
methods of science à la Lakatos (1970), the early Laudan (1977), or Worrall
(1988, 1989). Instead, it attempts to unearth the dynamics of changes in
methods and explain how epistemic agents change their criteria of theory
evaluation (Barseghyan, 2015, pp. 132-152). In addition, because scientonomy
has no associated normative theory, there is no worry of shoehorning the
historical episodes into the confines of some preferred methodology.

In the arrangement we are suggesting, historical narratives and general
scientonomic hypotheses are in a state of constant mutual adjustment. This
proposal is based on the notion of "epistemic iterations" developed by Chang
(2011, p. 114; see Chang, 2004, ch. 5, for a more detailed discussion):

> Epistemic iteration is a process in which we create successive stages of
> knowledge, each building on the preceding one, in order to enhance the
> achievement of certain epistemic goals... We do not start with
> indubitable facts, or unrevisable axioms. Instead, we start with a system
> of knowledge that we recognize as imperfect or even faulty, which is
> used for its own improvement. No fixed algorithm tells us how to
> proceed. But we have the impetus and constraints provided by the
> epistemic values and aims that we adopt.

Scientonomy originated from the acceptance of four laws and twenty-odd
theorems which collectively capture a variety of patterns found in the process
of scientific change, including compatibility, theory and method rejection,
splitting and merging of belief systems, underdetermination, role of
sociocultural factors, and more.[9] As fallibilists, we agree with Chang that these

---

[9] See https://www.scientowiki.com/Community:Scientonomy for the current state of the
theory.

laws and theorems are imperfect and revisable; the best we can do is to ensure their iterative improvement. In particular, scientonomy provides a set of accepted definitions, laws, and theorems, that can be applied to make sense of specific historical episodes. These historical episodes, in turn, often reveal problems with the theory thus leading to the refinement of its general hypotheses concerning the ontology and dynamics of scientific change. The theory informs the history and the history informs the theory. This process is iterated such that both our theoretical and historical knowledge gradually improve.

To facilitates this iterative advancement, we have developed a special *workflow*.[10] First, we document where we stand as a community in the encyclopedia of scientonomy. As we discover gaps, we document them as open questions for future research. Publications in our journal attempt to answer some of these questions by suggesting modifications to the current scientonomic theory. These suggested modifications are documented in the encyclopedia and subsequently evaluated by the community. If the community agrees that the suggested modification improves our knowledge, it becomes accepted and the respective pages of the encyclopedia are altered. If, however, the community doesn't find the modification acceptable or if there is no agreement among scientonomists, then our current theory remains intact.

This workflow allows us to circumvent the problem of cherry-picking. Cherry-picking arises because a single paper or book proposes a theory based on evidence that may be misrepresenting the more general population. It must be admitted that any first attempt at a theory will inevitably have this problem, because the conceivers will always have access to limited data. However, since in this workflow suggested modifications are evaluated by the community rather than individual authors and a few reviewers, this workflow incorporates the expertise of an entire (diverse) community during each iteration. In this evaluation process, any historical episode within this domain can become relevant; any and all counterexamples are entertained. This iterative process allows us to move gradually improve our theoretical knowledge as a result of *many iterations* with the historical record. As both historical data and theory improve, the risk of cherry-picking gradually diminishes.

This iterative process has been important for developing our theory. We have accepted new formulations of the law of theory acceptance (Patton, Overgaard, & Barseghyan, 2017), as well as a new formulation of the law of method employment (Sebastien, 2016). Our ontology of epistemic elements has

---

[10] See https://www.scientowiki.com/Scientonomic_Workflow for a detailed exposition of the workflow.

expanded from the initial ontology of theories and methods to the ontology of theories, methods, and questions (Rawleigh, 2018; Barseghyan, 2018). There are other suggested modifications that are currently being evaluated by the community (e.g. Barseghyan & Levesley, 2021).[11]

We should pause to note that, despite possible appearances, scientonomy does not conflict with the idea of the *disunity of science in most of its forms*. Ever since the landmark collection of Galison and Stump (1996), there has been an overwhelming flood of research aimed to show that science is, in some sense, a disunified phenomenon. The literature on the disunity of science has been, perhaps ironically, quite multifarious and has presented various ways in which science is disunified.

One common strategy of some proponents of the disunity of science is to show that, for a given notion, there is no single, coherent concept that unites the many instances of that notion. At best, there remains a family resemblance between instances of that notion. For example, while many areas of science purport to "explain" their domain of interest, there is little that unites these kinds of explanations. Some phenomena are explained via general laws, some are explained by causal means, others are explained by mechanisms, and some domains do not purport to "explain" anything (Mantzavinos, 2016; Khalifa, Doble, & Millson, 2020). Or, consider the following thesis (Psillos, 2006, p. 131):

> Most of the philosophical discussion about the metaphysics of causation has been dominated by what I call the 'straightjacket': the view that there is a single, unified and all-encompassing metaphysical story to be told as to what causation is. It has been presumed that the aim of philosophical inquiry is to tell this story... This paper questions the plausibility and fruitfulness of the 'straightjacket' as a whole. It lays out a number of ways to deny the straightjacket, ranging from some mild ones to some genuinely pluralistic. It outlines and defends a version of causal pluralism according to which causation is very much like the common cold: a rather loose condition with no single underlying nature.

The structure of this argument is common amongst proponents of different aspects of the disunity of science movement: concept *C* (in this case, causation) has "no single underlying structure" and this is demonstrated by investigation of a number of distinct uses of *C.* This kind of pluralism "applies widely to concepts,

---

[11] See https://www.scientowiki.com/Category:Modification for a complete list of suggested modifications.

explanations, virtues, goals, methods, models, and kinds of representations" (Cat, 2017) for different members of the disunity movement.

For others, there are many different "styles of reasoning" where different epistemic factors play differing roles depending on the phenomena being dealt with and the goal of the researcher (Hacking, 1996). In this sense of disunity, the various dimensions of scientific theorizing are unlikely to play the same epistemic roles in different contexts. As such, when we look at things from a more global perspective, we are forced to become pluralists about such notions.

For others, the disunity is primarily "metaphysical". What precisely this means varies from thinker to thinker. For some, this means that there are no general (or universal) laws of nature but there is only local patchwork of facts (Cartwright, 1999). For others, it means that some domains are autonomous and non-reducible to other domains, e.g., that psychological laws are not reducible to neuroscientific laws (Fodor, 1974). Some claim that there is no general structure to the world (Waters, 2017), and others think of disunity in the sense that individual entities are simultaneously parts of many natural kinds which are not reducible to each other (Dupré, 1993). These claims are often justified, at least partially, by an appeal to the commitments of specific scientific theories. As such, this claim is not so much about the disunity of science *itself* but more the claim that what the best science teaches us is that the world *itself* is disunified in some sense. As such, the metaphysical disunity of the world is an *interpretation* of the results of science rather than a claim about the process of scientific change itself.

While we do not want to defend any of these particular theses here, we do want to state that scientonomy is compatible with all of these families of views. Scientonomy admits that many aspects of scientific theorizing change throughout history and that different epistemic agents can employ different methods of theory evaluation or accept different notions of similar concepts. Despite the fact that scientonomy is a general theory of scientific change and may appear frightening to a proponent of the disunity of science at first glance, the two movements are not in conflict. Scientonomy does not presume any particular metaphysical theory nor do we swear allegiance to the content of particular scientific theories. In addition, our whole approach is based on the assumption of *the disunity of method*: criteria of theory evaluation can and often do vary drastically across epistemic agents, time periods, and fields of inquiry. More generally, scientonomy does not make any specific claims about the compositions of theories, methods, or practices of particular epistemic agents or what future compositions may look like. As such, they may well be exceptionally disunified and highly context-dependent.

The only kind of disunity scientonomy distinguishes itself from is *absolute* disunity: if there were no underlying mechanism of scientific change, if there were no general *patterns* of how science changes throughout time, then scientonomy would be impossible. Our contention is that there are general patterns of how theories and methods change. The disunity of science movement is grounded in the *content* of these theories and methods, which scientonomy is neutral towards. The type of unity that scientonomy assumes concerns the underlying *patterns* of how these theories and methods change. Insofar as a proponent of the disunity of science does not deny the existence of such underlying patterns, there is no conflict between our approaches.

For example, one general pattern that we have discovered manifests itself during transitions from one employed method to the next. Consider an agent who customarily tests the efficacy of drugs in *controlled* trials without blinding either the patients or researchers with respect to group allocation. Now, how would this agent react if she came to learn about the existence of the placebo effect? She would naturally alter her expectations and would require that a drug's efficacy be shown in a trial where the possible placebo effect is taken into account. Thus, she would change her method from her initial *controlled* trial method to a *blind* trial method. She would change her method again if she were to discover the experimenter's bias – this time from a *blind* trial method to a *double-blind* trial method that requires that both the patients and the researchers be blinded with respect to group allocation. What we have here is a pattern that is arguably instantiated in all cases of method employment. This pattern is succinctly captured in what is known as *the third law* of scientific change – the law of method employment: *a method becomes employed only when it is deducible from some subset of other employed methods and accepted theories of the time* (Barseghyan, 2015, pp. 132-152; Sebastien, 2016).[12]

There are innumerable instances of this law in action in different time periods and fields of inquiry. Consider, for example, the early 18th-century transition in natural philosophy from the method of intuition schooled by experience to the method of hypotheses. As shown by Feyerabend (1975, pp. 232-233) and McMullin (1988, pp. 32-34), the transition was in part a result of the community's acceptance of the view that the world as it appears in observations and experiments is a product of a potentially very *complex* underlying mechanism – a view that was implicit in the ideas of empiricists and rationalist alike. Take, for instance, Locke's distinction between primary and secondary qualities or Descartes's idea that the only attribute of matter is

---

[12] The idea that our knowledge shapes our criteria of theory evaluation is of course now new; it can be found in Kuhn (1970, p. 109), Feyerabend (1975, pp. 232-233), Shapere (1980), Laudan (1984, p. 39), McMullin (1988, pp. 32-34), and Brown (2001, pp. 137-140). Yet, what wasn't clear is the actual mechanism of this shaping: specifically, do employed methods *logically follow* from accepted theories, or do they merely *cohere* with accepted theories, or is there some other mechanism in play? The third law is scientonomy's attempt to fill this gap.

extension while color, smell, taste, and sound are the products of complex combinations and motions of these extended material parts. Once this fundamental idea became accepted, it led to changes in the method of theory evaluation; namely, it became acceptable to suggest explanatory hypotheses about inner mechanics of the world provided that they withstood observational tests. The method of hypotheses became employed, in full accordance with the third law, because it was a deductive consequence of the idea of underlying complexity, among other things (Barseghyan, 2015, pp. 145-150).

Illustrations of this pattern can be found as far back as ancient science. Consider, for instance, Plato's requirement that true knowledge is to be attained by means of reason alone. For example, Plato accepted the idea that each of the four terrestrial elements; earth, water, air, and fire, is composed of one of the Platonic solids – cube, icosahedron, octahedron, and tetrahedron respectively, on the basis of purely rational considerations (Zeyl & Sattler, 2017). Plato's requirement is clearly based on several assumptions, including the belief that truth is attainable, but that the senses are incapable of revealing the true reality. This is because the senses can only tell us about physical objects, which are merely imperfect imitations of eternal Platonic forms (Lindberg, 2008, pp. 35-37). Once again, what we have here is an employed method that is a deductive consequence of certain theories accepted by the agent.

Importantly, what we are "selling" here is not any specific theory, but the very idea of scientonomy as a viable descriptive science of scientific change, i.e., the idea that the process of scientific change exhibits some general patterns and that our workflow makes it possible to iteratively uncover these patterns. We may be inaccurate in our description of these patterns, or they may be more limited than we suppose. However, we have good reasons to suppose that future iterations of our theory will catch these problems at some point, thus improving our theory from where it stands today.

5. How can Normative Philosophy of Science Benefit from Descriptive Scientonomy?

Now, one may ask, this is all well and good, but what's in it for a philosopher of science? More specifically, how can a *normative* philosophy of science have any input from *descriptive* scientonomy, given the well-known issues with inferring "ought" from "is" (Giere, 2011; Schickore, 2011)? While we agree that no set of purely descriptive premises can warrant a normative conclusion, we hold that descriptive scientonomy can play an important role in normative philosophy of science.

From the logical point of view, a descriptive premise can have a normative conclusion if it is coupled with a normative premise. Take, for instance, the descriptive proposition "humans cannot survive without food for prolonged time periods". In and of itself, it cannot warrant an inference to the normative conclusion "starvation should be avoided". To make the inference logically sound, we must take the descriptive premise in conjunction with another normative premise, such as "what is deadly for humans should be avoided" (Figure 3.1). More generally, there is a way for descriptive theories to have input in normative considerations provided that one of the premises is itself normative (Pigden, 2010). Thus, there is a way for descriptive scientonomy to influence the normative philosophy of science without falling into the trap of the naturalistic fallacy.

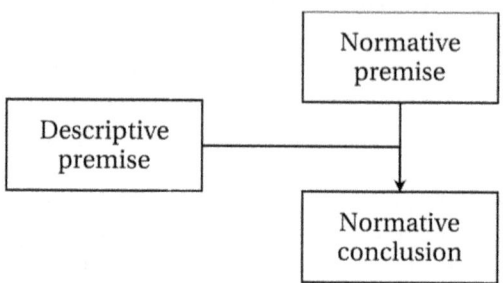

Figure 3.1: At least one of the premises of a normative conclusion must be normative

We envisage a relation between the two which is similar to that between clinical epidemiology and healthcare, between physics and engineering, or between descriptive economics and economic policy. In all these cases, the respective normative conclusions are drawn from a variety of descriptive and normative premises. When healthcare practitioners recommend a certain drug, therapy, or rehabilitation program, they draw on the descriptive findings of clinical epidemiology concerning the effects of these exposures. Yet, they conjoin these descriptive findings with the basic normative premise that a physician should strive for the improvement of the health of her patients. Similarly, when engineers argue that a certain structure should be demolished, they rely on certain physical theories to show that the structure has become unstable and can no longer be maintained. They do so by relying on the basic normative premise that what cannot be maintained should be demolished. It is our belief that the findings of descriptive scientonomy can play an analogous role in arguments for certain normative methodological conclusions.

For example, consider a hypothetical philosopher of science who believes that we should preserve the unity of an epistemic community where possible. According to the current scientonomic theory, when two currently unaccepted mutually incompatible theories simultaneously meet the requirements of a certain community, this results in a *mosaic split*, i.e., the community splits itself into two separate communities each accepting one of the incompatible theories (Barseghyan, 2015, pp. 202-216). Such splits are not uncommon in the history of science (e.g., the split between Newtonians and Cartesians during the first four decades of the 18th century). With this descriptive knowledge at hand, our philosopher can make a normative recommendation to ensure that the two contenders are evaluated such that it is impossible for both contenders to simultaneously pass the test. For instance, if our philosopher were interested in novel predictions, she could recommend testing only the *conflicting* predictions of the two contenders (Figure 3.2). While the soundness of this conclusion might be questionable, the validity of the argument is not, and that's all we wish to demonstrate here.

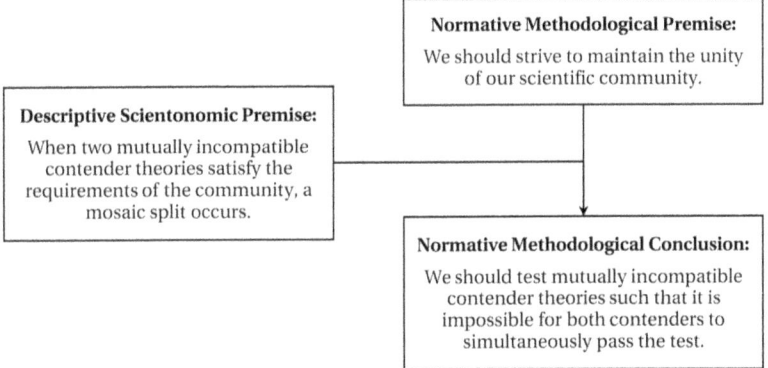

Figure 3.2: An example of a normative methodological conclusion drawn from the conjunction of a descriptive scientonomic premise and a normative methodological premise

It is also important to appreciate that descriptive theories *cannot determine* any normative conclusion by themselves. Thus, an agent making a decision is by no means bound by a descriptive theory, because the latter can never tell anyone what they should or should not do. However, it can tell the agent that, given what we know, their plan – if implemented – can have such-and-such consequences. At times, it can even tell them that the plan is unfeasible.

Consider an engineer who is planning to build a bridge according to the principles of orgonomy, the study of the so-called "orgone energy" or "orgone radiation". The engineer could, in principle, decide to go down that road. Yet, given our best physical theories, we can say that such a decision will likely result in a bridge that won't stand. Alternatively, imagine that our engineer decides to build their bridge in accordance with the laws of classical physics. As a descriptive theory, this physics cannot suggest that a bridge should be built in such-and-such a way, for there is a myriad of potential ways to build bridges which can stand equally well. Yet, it can tell the agent what would likely happen if they were to build it in a certain way. Thus, a descriptive theory can tell us that, according to the laws of the theory, certain bridge designs are questionable, while others are likely to produce a proper bridge. Or consider an agent who wants to build a perpetual motion machine. While the laws of thermodynamics cannot tell the agent not to try to build one, they can clearly indicate that, given what we know about entropy, the plan is not likely to succeed (Naess, 2005, pp. 71-84). Likewise, there is nothing in our anatomical theories that could prevent the likes of Baldrick to try and solve the problem of their mother's low ceiling by cutting her head off. What our anatomical theories can tell us is that such a plan will likely be fatal for Baldrick's mother. In short, our descriptive theories provide us with *can*-s, while our normative theories provide us with *ought*-s.

Similarly, we believe that descriptive scientonomy can help the philosopher of science evaluate how feasible different normative philosophical proposals are. Consider a methodologist who suggests that the community of clinical epidemiologists should prefer patient testimonies over randomized controlled trials when evaluating the therapeutic efficacy of a drug. Now, if the current third law of scientonomy is correct, we can predict that this methodological proposal is not viable. It follows from the third law, that any community that accepts the placebo effect and the experimenter's bias will only accept the therapeutic efficacy of a drug if it has been shown in such trials that account for the possible bias introduced by these phenomena. Thus, simply asking the patients what they think about the efficacy of the drug won't convince the community, as it doesn't take into account the possible placebo effect or the experimenter's bias.[13]

Yet, to reiterate, descriptive theories cannot by themselves *force* any normative decisions. To appreciate this, take the famous case of the invention of the barometer in the middle of the 17th century. According to the principles of the then-accepted Aristotelian natural philosophy, air was a light element with a natural tendency to ascend towards the periphery of the sublunar region; as such, it couldn't have any heaviness. Thus, it followed from the descriptive theory that the very idea of the barometer was utopian. Yet, this didn't stop the likes of Torricelli and Pascal who opted to proceed contrary to the descriptive knowledge of the time. The invention of the barometer became an anomaly which a few generations of Aristotelian natural philosophers had to grapple with until the eventual rejection of the whole Aristotelian framework circa 1700 (Brockliss, 2002).

---

[13] Of course, it might in principle turn out to be the case that alternative methods of patient testimony might circumvent the issues associated with the placebo effect and the experimenter's bias.

The same goes for the relation between descriptive scientonomy and normative philosophy of science. Suppose, for the sake of argument, that despite the best predictions of current scientonomy, clinical epidemiologists nevertheless decide to follow the recommendations of the above methodologist and start ranking the results of patient testimonies higher than those obtained in randomized controlled trials. This would be a serious anomaly for the current third law and would call for its modification.[14]

6. Conclusion

As we have shown, both the descriptive history of science and normative philosophy of science need some descriptive knowledge about the general patterns of scientific change. *Scientonomy* aims to provide precisely such a descriptive theory about the general mechanism that underlies changes in our theories about the world and our methods of theory evaluation. In the picture that we are suggesting, scientonomy plays an intermediary role between the descriptive history of science and the normative philosophy of science.

While descriptive scientonomy cannot make any normative claims on how science ought to be practiced, it can still benefit the philosophy of science. By taking the descriptive findings of scientonomy and coupling them with some additional normative premises, the philosopher of science can draw normative methodological conclusions which can inform and guide future scientific practices. Just as any other descriptive theory, the current scientonomic theory cannot necessitate one right way of doing things; instead, it can help the philosopher by suggesting that certain methodological prescriptions are more feasible than others. The relation that we are depicting here is not unique: it's essentially the same arrangement that exists between many descriptive fields of inquiry and their normative siblings. All it takes now is the good will of the HPS community to implement this proposal in practice.

Bibliography

Allchin, D. (2003). Lawson's Shoehorn, or Should the Philosophy of Science Be Rated 'X'? *Science & Education*, 12, 315-329.

Bar-Hillel, Y. (1968). The Acceptance Syndrome. In Lakatos (Ed.) (1968), 150-161.

Barseghyan, H. (2015). *The Laws of Scientific Change.* Springer.

Barseghyan, H. (2018). Redrafting the Ontology of Scientific Change. *Scientonomy*, 2, 13-38.

Barseghyan, H. & Levesley, N. (2021). Question Dynamics. *Scientonomy*, 4, 1-19.

---

[14] This touches upon an important question that is beyond the scope of this paper: how can normative philosophy of science have input for scientonomy?

Baumrin, B. (Ed.) (1963). *Philosophy of Science: The Delaware Seminar Volume 2*. Interscience Publishers.

Brockliss, L. (2002). Harvey, Torricelli and the Institutionalization of New Ideas in the 17th Century France. In Detel & Zittel (Eds.) (2002), 115-134.

Brown, J. R. (2001). *Who Rules in Science?* Harvard University Press.

Bueno, O. (2008). Structural Realism, Scientific Change, and Partial Structures. *Studia Logica*, 89, 213-235.

Burian, R. (2001). The Dilemma of Case-studies Resolved: The Virtues of Using Case-studies in the History and Philosophy of Science. *Perspectives on Science*, 9, 383-404.

Carnap, R. (1966). Probability and Content Measure. In Feyerabend & Maxwell (Eds.) (1966), 248-260.

Cartwright, N. (1999). *The Dappled World: A Study of the Boundaries of Science*. Cambridge University Press.

Cat, J. (2017). The Unity of Science. In Zalta, E. N. (Ed.) (2021). *The Stanford Encyclopedia of Philosophy (Fall 2021 Edition)*. Retrieved from: https://plato.stanford.edu/archives/fall2021/entries/scientific-unity/.

Chang, H. (1999). History and Philosophy of Science as a Continuation of Science by Other Means. *Science & Education*, 8(4), 413-425.

Chang, H. (2004). *Inventing Temperature: Measurement and Scientific Progress*. Oxford University Press.

Chang, H. (2011). Beyond Case-Studies: History as Philosophy. In Mauskopf & Schamltz (Eds.) (2011), 109-124.

Chang, H. (2012). *Is Water H_2O? Evidence, Realism and Pluralism*. Springer.

Detel, W. & Zittel, C. (Eds.) (2002). *Ideals and Cultures of Knowledge in Early Modern Europe*. De Gruyter.

Domski, M. & Dickson, M. (Eds.) (2010). *Discourse on a New Method: Reinvigorating the Marriage of History and Philosophy of Science*. Open Court.

Donovan, A., Laudan, L., & Laudan, R. (Eds.) (1992). *Scrutinizing Science. Empirical Studies of Scientific Change*. The Johns Hopkins University Press.

Dupré, J. (1993). *The Disorder of Things: Metaphysical Foundations of the Disunity of Science*. Harvard University Press.

Feyerabend, P. (1963). How to Be a Good Empiricist: A Plea for Tolerance in Matters Epistemological. In Baumrin (Ed.) (1963), 3-39.

Feyerabend, P. (1970). Consolations for the Specialist. In Feyerabend (1981), 131-162.

Feyerabend, P. (1975). *Against Method*. New Left Books.

Feyerabend, P. (1981). *Problems of Empiricism: Philosophy Papers Volume II*. Cambridge University Press.

Feyerabend, P. & Maxwell, G. (Eds.) (1966). *Mind, Matter and Method: Essays in Philosophy and Science in Honor of Herbert Feigl*. University of Minnesota Press.

Fodor, J. (1974). Special Sciences (Or: The Disunity of Science as a Working Hypothesis). *Synthese*, 28(2), 97-115.

Galison, P. & Stump, D. (Eds). (1996). *The Disunity of Science: Boundaries, Contexts, and Power*. Stanford University Press.

Giere, R.N. (2011). History and Philosophy of Science: Thirty-Five Years Later. In Mauskopf & Schamltz (Eds.) (2011), 59-65.

Hacking, I. (1996). The Disunities of the Sciences. In Galison & Stump (Eds.) (1996), 37-74.

Herring, E., Jones, K. M., Kiprijanov, K. S., & Sellers, L. M. (Eds.) (2019). *The Past, Present, and Future of Integrated History and Philosophy of Science*. Routledge.

Home, R. W. (2003). Mechanics and Experimental Physics. In Porter (Ed.) (2003), 354-374.

Khalifa, K., Doble, G., & Millson, J. (2020). Counterfactuals and Explanatory Pluralism. *The British Journal for the Philosophy of Science*, 71 (4), 1439-1460.

Kuhn, T. (1970). Reflections on My Critics. In Lakatos & Musgrave (Eds.) (1970), 231-278.

Kuhn, T. (1977). *Essential Tension: Selected Studies in Scientific Tradition and Change*. University of Chicago Press.

Kuukkanen, J.-M. (2017). Lakatosian Rational Reconstruction Updated. *International Studies in the Philosophy of Science*, 31(1), 83-102.

Lakatos, I. (1968). Changes in the Problem of Inductive Logic. *Studies in Logic and the Foundations of Mathematics*, 51, 315-417.

Lakatos, I. (1970). Falsification and the Methodology of Scientific Research Programmes. In Lakatos (1978), 8-101.

Lakatos, I. (1971). History of Science and its Rational Reconstructions. In Lakatos (1978), 102-138.

Lakatos, I. (1978). *Philosophical Papers, Volume I*. Cambridge University Press.

Lakatos, I. (Ed.) (1968). *The Problem of Inductive Logic*. North Holland Publishing Company.

Lakatos, I. & Musgrave, A. (Eds.) (1970). *Criticism and the Growth of Knowledge*. Cambridge University Press.

Lakatos, I. & Zahar, E. (1976). Why did Copernicus's Research Programme Supersede Ptolemy's? In Lakatos (1978), 168-192.

Laudan, L. (1977). *Progress and its Problems. Toward a Theory of Scientific Growth*. University of California Press.

Laudan, L. (1984). *Science and Values*. University of California Press.

Laudan, L. (1986). Some Problems Facing Intuitions Meta-Methodologies. *Synthese*, 67(1), 115-129.

Lindberg, D. (2008). *The Beginnings of Western Science*. The University of Chicago Press.

Mantzavinos, C. (2016). *Explanatory Pluralism*. Cambridge University Press.

Mauskopf, S. & Schamltz, T. (Eds.) (2011). *Integrating History and Philosophy of Science: Problems and Prospects*. Springer.

McMullin, E. (1988). The Shaping of Scientific Rationality: Construction and Constraint. In McMullin (Ed.) (1988), 1-47.

McMullin, E. (Ed.) (1988). *Construction and Constraint. The Shaping of Scientific Rationality*. University of Notre Dame Press.

Mitchell, S. (2009). *Unsimple Truths: Science, Complexity, and Policy*. University of Chicago Press.

Mizrahi, M. (2015). Historical Inductions: New Cherries, Same Old Cherry-Picking. *International Studies in the Philosophy of Science*, 29(2), 129-148.

Naess, A. (2005). *The Pluralist and Possibilist Aspect of the Scientific Enterprise.* Springer.

Nickles, T. (Ed.) (1980). *Scientific Discovery, Logic and Rationality.* D. Reidel.

Nickles, T. (1995). Philosophy of Science and History of Science. *Osiris*, 10, 139-163.

Patton, P., Overgaard, N., & Barseghyan, H. (2017). Reformulating the Second Law. *Scientonomy*, 1, 29-39.

Pigden, C. (2010). On the Triviality of Hume's Law: A Reply to Gerhard Schurz. In Pigden (Ed.) (2010), 217-238.

Pigden, C. (Ed.) (2010). *Hume on Is and Ought.* Palgrave Macmillan.

Pitt, J. C. (2001). The Dilemma of Case-studies: Toward a Heraclitian Philosophy of Science. *Perspectives on Science* 9(4): 373-382.

Popper, K. (2002). *The Logic of Scientific Discovery.* Routledge.

Porter, R. (Ed.) (2003). *The Cambridge History of Science. Volume 4: Eighteenth-Century Science.* Cambridge University Press

Psillos, S. (2006). Causal Pluralism. In Vanderbeeken & D'Hooghe (Eds.) (2006), 131-151.

Rawleigh, W. (2018). The Status of Questions in the Ontology of Scientific Change. *Scientonomy*, 2, 1-12.

Reichenbach, H. (1938). *Experience and Prediction: An Analysis of the Foundations and the Structure of Knowledge.* University of Chicago Press.

Schickore, J. (2011). More Thoughts on HPS: Another 20 Years Later. *Perspectives on Science* 19(4), 453-481.

Sebastien, Z. (2016). The Status of Normative Propositions in the Theory of Scientific Change. *Scientonomy*, 1, 1-9.

Shapere, D. (1977). What Can the Theory of Knowledge Learn from the History of Knowledge? *The Monist*, 60, 488-508.

Shapere, D. (1980). The Character of Scientific Change. In Nickles (Ed.) (1980), 61-116.

Slater M. & Yudell, Z. (Eds.) (2017). *Metaphysics and the Philosophy of Science: New Essays.* Oxford University Press.

Suppes, P. (1978). The Plurality of Science. *PSA: Proceedings of the Biennial Meeting of the Philosophy of Science Association*, 1978, 3-16.

Vanderbeeken, R. & D'Hooghe, B. (Eds.) (2006). *Worldviews, Science and Us: Studies of Analytical Metaphysics: A Selection of Topics from a Methodological Perspective.* World Scientific Pub Co Inc.

Waters, K. (2017). No General Structure. In Slater & Yudell (Eds.) (2017), 81-108.

Wilson, C. (2003). Astronomy and Cosmology. In Porter (Ed.) (2003), 328-353.

Worrall, J. (1988). Review: The Value of a Fixed Methodology. *The British Journal for the Philosophy of Science*, 39, 263-275.

Worrall, J. (1989). Fix It and be Damned: A Reply to Laudan. *The British Journal for the Philosophy of Science*, 40, 376-388.

Wykstra, S. J. (1980). Toward a Historical Meta-Method for Assessing Normative Methodologies: Rationability, Serendipity, and the Robinson

Crusoe Fallacy. *PSA: Proceedings of the Biennial Meeting of the Philosophy of Science Association,* 1980, 211-222.

Zeyl, D. & Sattler, B. (2017). Plato's Timaeus. In Zalta, E. N. (Ed.) (2017). *The Stanford Encyclopedia of Philosophy (Winter 2017 Edition).* Retrieved from: https://plato.stanford.edu/archives/win2017/entries/plato-timaeus/.

Chapter 4

Integrating Scientonomy
with Scientometrics

Karen Yan

National Yang Ming Chiao Tung University

Meng-Li Tsai

National Ilan University

Tsung-Ren Huang

National Taiwan University

Abstract: Scientonomy is the field that aims to develop a descriptive theory of the actual process of scientific change (Barseghyan, 2015). Scientometrics is the field that aims to employ statistical methods to investigate the quantitative features of scientific research, especially the impact of scientific articles and the significance of scientific citations (Leydesdorff & Milojević, 2013). In this paper, we aim to illustrate how to methodologically integrate scientonomy with scientometrics to investigate both qualitative and quantitative changes of a scientific community. We will use a case study to achieve our aim. The case study is about a scientific community studying a physiological phenomenon called heart-rate variability. Moreover, we will argue that this methodological integration outperforms cases in which researchers only employ the resources from one of the two fields.

Keywords: scientonomy; scientometrics; heart-rate variability; GROBID; spaCy

***

1. Introduction

In 1946, the University of Melbourne started one of the first History and Philosophy of Science (HPS) programs in the world.[1] In the 1960s, HPS became a well-received academic discipline, a discipline that aspired to 'integrate' both the philosophy of science (PS) and the history of science (HS) to study scientific change. This aspiration of HPS is well captured by Lakatos's famous dictum that "philosophy of science without history of science is empty; history of science without philosophy of science is blind" (Lakatos, 1971, p. 102). Put differently, historians must start with a taxonomy about scientific change to perform their historical analyses. But which taxonomy to use is a philosophical decision that requires some philosophical work to decide. Historians also frequently rely on some implicit generalizations about particulars, events, or how they interact with each other to interpret the significance of the relevant historical materials. But it takes some philosophical work to justify why the employed generalizations hold. On the other hand, for philosophers to produce historically justifiable taxonomies and generalizations about scientific change, philosophers must work with actual historical evidence of scientific change. But, to what extent has this aspiration of integration been achieved?

Thomas Kuhn was recognized as the most influential figure of HPS during the 20th century. However, even Kuhn (1977, p. 5) admitted that while HS and PS both investigate scientific change, they aim at different goals. Whereas HS aims at producing historical narratives of particulars and events, PS aims at producing universal or lawlike generalizations.[2] Because HS and PS have different goals, Kuhn thought that practicing HS and PS at the same time, was like perceiving the duck-rabbit illusion. One can practice HS and PS separately, but one cannot practice HS and PS at the same time. If so, the notion of integrating two distinct entities into one coherent one does not apply to HPS. Kuhn thought the best-case scenario is to switch between HS and PS while investigating scientific change.

Unfortunately, the subsequent development of HPS didn't even achieve Kuhn's switching interpretation of how HS and PS can be integrated. As many scholars have pointed out, in HPS, HS and PS are often two distinct sets of

---

[1] https://arts.unimelb.edu.au/school-of-historical-and-philosophical-studies/study/history-and-philosophy-of-science

[2] It is worth noting that the notion of universal or lawlike generalization has been criticized severely by contemporary philosophers of science since the 1980s (Cartwright, 1983; Dupré, 1993; Bechtel & Richardson, 1993; Cartwright, 1999; Wimsatt, 2007; Mitchell, 2012). It is no longer accurate to say the goal of PS is to produce universal or lawlike generalizations. Perhaps it is more appropriate to say that PS aims to produce contextualized generalizations (Massimi, 2016).

practices and their research activities rarely have substantive overlaps. Researchers have not successfully demonstrated how to switch between the two while addressing the same question, and thereby achieve Kuhn's version of integration (Zammito, 2004; Golinski, 2012; Miller, 2012; Caneva, 2012).

This led many scholars to advocate restructuring how HS and PS are currently practiced in order to truly live up to the original aspiration of HPS. Integrated History and Philosophy of Science (IHPS) thus became a vibrant movement with the goal of seeking a fruitful and successful way to integrate HS and PS (Mauskopf & Schmaltz, 2011; Herring et al. (Eds.), 2019). Under this movement, a distinct field called scientonomy has been established by a group of scholars (Barseghyan, 2015; Rupik, 2019). Scientonomy is the field that aims to develop a descriptive theory of the actual process of scientific change. Scientonomy seeks to correct the tendency to downplay descriptive historical work on scientific change in PS and the tendency to ignore the need for a descriptive theoretical framework for doing historical analyses in HS. To achieve this aim, Barseghyan (2015) develops a tentative conceptual framework and historical hypotheses. More importantly, scientonomists have established a community environment that allows scholars, ideally from both HS and PS, to propose, revise, or reject elements in the tentative conceptual framework and historical hypotheses through an online platform.[3] One significance of this implementation is that it builds an environment in which HS and PS scholars can work in an overlapping way while each group can still focus on their respective research goals. Given the above innovative conceptual and community designs, scientonomists aspire to establish an "empirical science of science" that is a version of IHPS (Rupik, 2019).

We support the aspirations of scientonomy, and would like to suggest some methodological innovations to open up the possibilities to integrate HS, PS, and scientometrics into the study of scientific change. Scientometrics is the field that employs statistical methods to investigate the quantitative features of scientific research, especially the impact of scientific articles and the significance of scientific citations (Leydesdorff & Milojević, 2013). Some of the statistical methods can be applied to investigate scientific change in a quantitative way. In this paper, we aim to argue that it is fruitful to incorporate methods from scientometrics or involve scientometrists in the Scientonomy Community. We will substantiate our argument by illustrating how to integrate scientonomy and scientometrics methodologically. To that end, the paper will proceed as follows. In Sect. 2, we will present our case study and how we had employed some statistical methods to analyze the case study. In Sect. 3, we will summarize the details of the third law from scientonomy. In Sect. 4, we will

---

[3] https://www.scientojournal.com/

apply the third law to further analyze our case study based on our previous quantitative and qualitative results. In Sect. 5, we will further apply some statistical methods to empirically examine one result of applying the third law to our case study. In Sect. 6, we will show why our integrated method outperforms cases in which researchers only employ resources from one of the two fields.

2. HRV Case Study

Our case study concerns a scientific community studying a physiological phenomenon called heart-rate variability (HRV) and its clinical applications. HRV refers to "the oscillation in the interval between consecutive heartbeats as well as the oscillations between consecutive instantaneous heart rates" (Task Force of the European Society of Cardiology and the North American Society of Pacing and Electrophysiology, 1996, p. 354).[4] Hon and Lee (1963) first reported the clinical relevance of HRV for investigating fetal distress. Later, scientists recognized the potential of measuring HRV to observe how the sympathetic and parasympathetic nervous systems interact with each other (Rajendra Acharya et al., 2006, p. 1031). Since then, scientists have been working on developing HRV as a quantitative marker of autonomic activity to predict cardiovascular and other types of diseases (Kleiger et al., 1987; Malik et al., 1989; Bigger et al., 1992). After 2001, however, researchers focused more on HRV's applications in topics such as stress, sleep, and exercise (Yan, Tsai, & Huang, 2020).

Yan, Tsai, and Huang (2020) analyzed 19,795 articles published between 1970 and 2016 from the HRV community. In this work, part of our analysis involved three quantitative methods: citation analysis, classification analysis, and text analysis. Citation analysis and text analysis are part of the methods used in scientometrics. Our quantitative analyses generated two sets of quantitative and qualitative descriptions of how the HRV field changed over time between 1970 and 2016 (Yan, Tsai, & Huang 2020, figures 2 and 3). Here is a summary of our results. We used citation analysis to identify the most cited article (6460 times) in the field between 1970 and 2006 and name this article RA-11 (RA stands for review article and 11 is the article's assigned number). RA-11 was published in 1996 by the Task Force of the European Society of Cardiology and the North American Society of Pacing and Electrophysiology. We used classification analysis to classify all the articles into the following three categories: (A) foundational research on relevant mechanisms or technological advances, (B) correlation to cardiovascular disease, and (C) correlation to non-cardiovascular disease. The results of our classification analysis showed that,

---

[4] We will use 'Task Force 1996' in subsequent citations to this article.

between 1996 and 2001, the field changed its dominant research category: Category A and B were decreasing and category C was increasing. After 2001, category C research was consistently maintaining above 60% of the research done in the HRV field. This change of research categories is correlated with the publication year of RA-11, i.e., 1996. We then performed text analysis to calculate the frequency of each word in all the titles of the 19,795 articles used in our classification analysis. The results of our text analysis show that, between 1996 and 2001, sleep started showing up. After 2001, exercise, sleep, and stress became the three most frequent words. This shows that, after RA-11 was published in 1996, the HRV field shifted its research trend toward sleep, exercise, and stress, which are category C research. This shift is consistent with the data from our classification analysis. In short, the data from our citation analysis shows that RA-11 is the most cited article in the field. The data from our classification analysis shows that, after 1996, category C research gradually increased and consistently maintained above 60% of the research after 2001. The data from our text analysis shows a pattern of changes that is consistent with the data from our classification analysis. Based on these three sets of quantitative data, we can justify using RA-11 as the starting point of our case-based research and investigate the following research question (which is also generated from the three datasets): Why did the HRV field exhibit the identified quantitative and qualitative changes after RA-11 was published in 1996?

The above quantitative analyses led us to perform a detailed practice-based analysis of the content of RA-11 to collect more information regarding the correlation between RA-11 and the identified patterns of changes in the HRV field. Our practice-based analysis is inspired by the recent practice-turn in the philosophy of science. The idea is to use scientific practices as units of analysis for our philosophical investigation. We construe scientific practices as activities that are performed by scientists in order to achieve certain goals (Chang, 2011). In other words, scientific practice is conceptualized as a three-place relationship among scientists, their activities, and their goals. We applied a practice-based analysis to the content of RA-11 and found that the main goals of the task force that wrote it were to standardize measurement practices, recording practices, signal-processing practices, conceptual practices, and procedures and methods for measuring HRV for the two clinical applications. Moreover, they pointed to various possibilities of future clinical applications, specifically dealing with sleep and exercise (Task Force, 1996, p. 347). Thus, our practice-based analysis revealed that the task force that published RA-11 achieved various types of standardizing work for the HRV field.

Above, we observed that the HRV field underwent significant quantitative and qualitative changes since the publication of RA-11 in 1996. The major epistemic achievement of RA-11 was to standardize practices. Given all of that, how do we understand the relationship between the standardization accomplished by RA-11 and the subsequent changes in the field?

Here is where we see the opportunity to integrate the scientometric results with the conceptual framework provided by scientonomy. The quantitative results help us identify an actual pattern of changes in the HRV community based on 19,795 articles published between 1970 and 2016. Our practice-based analysis helps us to specify the exact question to further investigate philosophically. In the remainder of the paper, we will illustrate how to use the conceptual framework of scientonomy to help us deepen our understanding of the relationship between the achievement of RA-11 in establishing standards and the identified patterns of change in the HRV community. Moreover, we will also illustrate how to employ the conceptual tools of scientonomy and the quantitative tools of scientometrics together to produce a deeper understanding of the changes in the HRV field.

3. The Third Law of Scientific Change

Barseghyan (2015) proposed the theory of scientific change (TSC) as a descriptive theory of the actual process of scientific change. TSC is formulated in axiomatic deductive form and currently involves four laws. Among them, the third law (a.k.a. law of method employment) is proposed to capture how our "accepted theories" (i.e., accepted knowledge or propositions about the world) (Barseghyan, 2015, p. 3) shape employed methods. Barseghyan formulates the third law as follows (2015, p. 143):

> A method becomes employed only when it is deducible from other employed methods and accepted theories of the time, i.e. either (1) when it strictly follows from the other employed methods and accepted theories, or (2) when it implements some abstract requirements of other employed methods.

Barseghyan specifies two scenarios that can satisfy the deductibility requirement in the third law. One involves the standard notion of deduction in logic. The other involves the implementation relation between two things. Barseghyan uses the example of testing a drug's efficacy to illustrate the above two scenarios.

In the first scenario, suppose some scientists aim to design experiments to test the efficacy of a drug because they hold that the drug's effect must be confirmed experimentally. At the same time, their accepted theory contains two other pieces of knowledge. One is that medical conditions can be improved by factors other than the tested drug's effect. The other is that the controlled trial method can be used to test for such confounding factors in an experiment. Consequently, they deduce that the drug's effect on health has to be confirmed by the controlled trial method. In this case, some employed method (i.e., accept the results of a controlled trial) follows strictly from the other employed methods (i.e., accept results that have been confirmed by experiment) and accepted theories (i.e., the two pieces of knowledge about the world).

In the second scenario, suppose some scientists aim to employ the controlled trial method to test a drug's efficacy. However, they also know about the placebo effect, i.e., that patients receiving 'fake' treatment still often show improvement in their medical condition for psychosomatic reasons. If the placebo effect is not taken into account, any interpretive claim based on the experimental results will not be taken as the best available claim about the drug's efficacy. Consequently, the following abstract requirement is deduced: "When assessing a drug's efficacy, the possible placebo effect must be taken into account" (Barseghyan, 2015, p. 137). Historically, the blind trial method is designed to take the possible placebo effect into account. However, as Barseghyan emphasizes, an abstract requirement is not the same thing as a concrete method. A concrete method can implement an abstract requirement, but not vice versa. Moreover, many different concrete methods can implement the same abstract requirement. In this case, some employed method (i.e., the blind trial method) implements an abstract requirement of the method (i.e., when assessing a drug's efficacy, the possible placebo effect must be taken into account), and the abstract requirement is what was deduced from other employed methods (i.e., the controlled trial method) and accepted theories (i.e., taking the placebo effect into account).

For the sake of simplicity, we will simplify the third law as follows:

1. *M* is logically deduced from *A* (*M* stands for methods becoming employed and *A* stands for other employed methods and accepted theories of the times)

2. *M* implements *B* (*B* stands for some abstract requirements of other employed methods)

4. Applying the Third Law of Scientonomy to Analyze the HRV Case

In this section, we will illustrate how to apply the third law to analyze our HRV case. Our targeted question in Section 2 is: How do we understand the relationship between the standardization achieved by RA-11 and the subsequent changes in the field? We can apply the third law to reformulate this question as follows: What is the status of the standardization achieved by RA-11 if we analyze it from the perspective of the third law?

From now on, we will use S to stand for the standardization achievement of RA-11. By applying the third law to analyze S, we consider the following four interpretive possibilities:

(P1) S is M in (1)

(P2) S is A in (1)

(P3) S is M in (2)

(P4) S is B in (2)

In the following, we will analyze each interpretive possibility to examine whether or not it is consistent with our previous quantitative and qualitative data (Yan, Tsai, & Huang, 2020).

4.1 First Interpretive Possibility

(P1) means that the standardization achievement of RA-11 amounts to deducing a set of concrete methods from other employed methods and accepted theories of the time. (P1) seems to conflict with how a task force usually operates (Djulbegovic & Guyatt, 2019). The task force that published RA-11 is an expert group engaging in the process of reviewing the extant literature and formulating the most reasonable choice of standards as they see fit. Though the HRV experts in the task force certainly perform logical inference, it seems unlikely that they establish S only through logical deduction. As the task force stated: "The standards and proposals offered in this text should not limit further development but, rather, should allow appropriate comparisons, promote circumspect interpretations, and lead to further progress in the field". This passage shows that the task force does not take S to result from strict logical inferences with logical certainty. Instead, they take S to be a reasonable starting point that is open for reinterpretation and revision as long as it can further the development of scientific practices in the HRV field.

4.2 Second Interpretive Possibility

(P2) means that the standardization achievement of RA-11 amounts to identifying some already employed methods and accepted theories of the time. (P2) also seems to conflict with how a task force usually operates. For (P2) to be the case, then the nature of S is nothing but the compilation and description of the employed methods and accepted theories of the time. This interpretation seems rather unlikely. For one thing, (P2) has the potential to impugn the ingenuity that went into the task force's work. Even if it were the case that the task force is just compiling information regarding employed methods and accepted theories of time, the way in which they organized this information can be said to be innovative. No one had compiled this information before the task force, and it did take much effort to put every piece of information into a coherent whole. The task force also made various suggestions regarding measurement and recording practices that needed further study. They also suggested potential clinical applications worth pursuing. These achievements go far beyond merely compiling the relevant information regarding employed methods and accepted theories of time. Thus, if S was innovative and added something new to the HRV community at the time, S cannot be just a set of employed methods and accepted theories of the time.

4.3 Third Interpretive Possibility

(P3) means that the standardization achievement of RA-11 amounts to designing a set of concrete methods that implements some abstract requirement. (P3) also does not seem consistent with the content of RA-11. At various places in RA-11, the task force stated their suggestions or recommendations in the following format: When employing this particular measuring or recording method/procedure/equipment, one should do (or not do) some action or one should follow some requirement (Task Force, 1996, pp. 357, 359-360, 364-365, 370-371). Though RA-11 did compile information regarding various measuring and recording practices, related experimental pieces of equipment, and clinical applications, the task force did not design the relevant methods to implement some abstract requirement as someone designing the blind trial method might do in order to implement the abstract requirement for taking the placebo effect into account. It is more likely the case that the task force compiled various methodological, technological, and biological information regarding HRV, and that they made suggestions and recommendations about how to employ the relevant methods, equipment, and information.

4.4 Fourth Interpretive Possibility

(P4) means that the standardization achievement of RA-11 amounts to establishing a set of abstract requirements. We think (P4) gives the best possible interpretation of our HRV case, out of the four interpretive possibilities generated by applying the third law. The task force repeatedly made various suggestions and recommendations regarding how to employ various types of HRV research practices: What should be done and what should not be done, how to use relevant technological equipment, and where to look for future possibilities for clinical applications.

However, to further substantiate (P4), some empirical data is needed. It does not make sense to interpret S as a set of abstract requirements if no members of the HRV community actually design concrete methods to implement some part of S. Here is where we see another opportunity to integrate scientometrics with scientonomy. We aim to illustrate how to apply some scientometric tools to investigate (P4) empirically and gather some empirical support for (P4).

It is worth noting that Barseghyan (2015, p. 151) has argued that "devising a new method that would implement abstract requirements takes a fair amount of ingenuity and, therefore, there are no guarantees that these abstract requirements will be immediately followed by a new concrete method". Thus, it is acceptable that the establishment of the abstract requirement and the implementation of the abstract requirement do not happen at the same time.

5. Employ Scientometric Tools to Examine (P4) Generated from Scientonomy

In this section, we will report a set of quantitative analyses. We performed them in order to empirically examine (P4). We aim to prove that S did play the role of establishing a set of abstract requirements for the HRV community to implement in their research.

5.1 Literature Search and Analyzed Samples

We quantified how RA-11 influenced later research. As a case study, we used Web of Science[5] to track citations of RA-11 between 1997 and 2019. Our literature search on Web of Science found 8,722 articles that cited RA-11 as of April 20, 2020.

To analyze the contents of the aforementioned body of literature, we had to obtain the PDF files of these articles. For this step, we leveraged the "Find Full Text" feature of the reference manager EndNote to automatically download PDF files of the literature, if available. In the end, only 4,828 PDF files were found and subjected to our later analysis.

5.2 Analysis Methods & Results

The goal of our scientometric analysis was to identify whether RA-11 played the role of establishing a set of abstract requirements for later HRV articles. Specifically, we aimed to determine how many times, and in which section, an article cited RA-11. Though seemingly simple, this goal was technically challenging in terms of text parsing because the downloaded 4,828 articles were published in 100 different journals, each of which had its own article formats and subscribed to various citation styles. In favor of a solution that is generalizable to the analysis of a different body of literature, we leveraged the bibliographic analyzer GROBID to automatically convert each PDF file into an

---

[5] http://webofknowledge.com/

XML file that shared the same set of tagged article elements (e.g., headers and references) across different articles/journals (Lopez, 2009). As a result, we could programmatically locate in-text citation texts and their corresponding headers or subheaders.

Note that the headers and especially subheaders of an article can be named freely by the authors. For example, even though RA-11 was cited three times under the section "3. Results" by Cicone and Wu (2017), these citations were there to support their methodological discussions in the subsections of "3.1. Quantities for evaluation" and "3.2. Simulated Database". Therefore, it was also technically challenging to classify each in-text citation of RA-11 into a conventional article section, namely "Introduction", "Methods", "Results", "Discussion", or "Conclusions". We implement an imperfect and yet effective solution by automatically categorizing each subheader into a conventional article section based on their semantic similarity. Here we used the natural language processor spaCy to convert each subheader into a sentence vector, which was embedded in a space where distance represents semantic dissimilarity (Honnibal & Montani, 2017). Then, the sentence vector of each header/subheader was compared to that of each conventional article section in terms of their cosine similarity, and only a cosine similarity larger than .5 was treated as a successful attempt of section categorization.

Combining all of the advanced text analysis techniques mentioned above, in the end, our analysis program managed to successfully analyze 2,240 out of 4,828 articles where success for each analyzed article was operationally defined as finding at least one in-text citation of the review paper and semantically categorizing each subsection title (i.e., subheader) into a conventional article section (Table 4.1).

	1997~ 1999	2000~ 2004	2005~ 2009	2010~ 2014	2015~ 2019
S1: Web of Science	479	1264	1720	2120	3139
S2: EndNote PDFs	218	664	833	1215	1898
S3: Successfully Analyzed	58	222	331	612	1017

Table 4.1: The number of articles in each stage (row) grouped by years (column)

Our result shows that RA-11 is cited in a method section 349 times (Table 4.2).

Introduction/ Background	Method/ Methods	Result/ Results	Discussion/ Discussions	Conclusion/ Conclusions
1911	349	74	527	263

Table 4.2: Number of RA-11 in-text citation in each section

It should be pointed out that, strictly speaking, the 2,240 articles are not representative samples of the original 8,722 articles (Table 4.2). This is because the results of our scientometric analysis are slightly systematically biased by the limitations of the informatics tools we employed.[6] However, we expect that, as informational techniques develop in the future, we will be able to automatically download all the PDF files and extract all the articles. Consequently, the problem of sampling bias in our method will be eliminated. Moreover, in this section, our goal is to show how scientonomists can employ our scientometric analysis to gather some empirical support for (P4). Our result at least shows that there are a good number of members of the HRV community citing RA-11 for methodological reasons (349 times). Assuming the standard practice of writing a scientific research paper, the method section is the place where authors specify how they design and execute their experiments. As we have argued before, the standardization achievement of RA-11 is best interpreted as a set of abstract requirements, thus the act of citing RA-11 in the method section is best interpreted as implementing some abstract requirements from RA-11. The following are four quoted examples from articles citing RA-11 at the method section:

HRV analysis was performed in the frequency domain (PSA) on each of three 5-minute ECG recordings (baseline, active standing, paced breathing), in accordance with the guidelines of the Task Force of the European Society of Cardiology and the North American Society of Pacing and Electrophysiology (Nicolini et al., 2014, p. 4);

HRV was analyzed using a computer program based on commercially available software (LabVIEW, National Instruments, Austin, TX, U.S.A.), which meets the recently published criteria of HRV measurement, [i.e. those from RA-11] (Keyl et al., 1997, p. 337);

Time-domain variables were evaluated in accordance with the guidelines of the Task Force of the European Society of Cardiology and the North American Society of Pacing and Electrophysiology (Weissman et al., 2009, p. 119);

Moreover, the experimental procedures were performed between 1 and 4 P.M. to standardize the protocol [according to RA-11] (Raimundo et al., 2013, p. 489).

---

[6] The χ^2 tests for independence between the S1 & S2 (χ^2=41.38; p=.001), S2 & S3 (χ^2=75.68; p=.001), S1 & S3 (χ^2=158.08; p=.001) were significant.

These four examples illustrate how HRV researchers perform their experiments in ways that implement some abstract requirements from RA-11. Since the practice of writing scientific journal articles is standardized, it is reasonable to expect that the other 345 examples of citing RA-11 at the method section are similar to the above four examples.

In short, in this section, we have employed some advanced text analysis techniques to illustrate how to further integrate scientonomy with scientometrics. We showed that our scientometric results can provide some empirical support for (P4).

6. Why the Integrated Approach Outperforms

At this point, let's revisit Lakatos's famous dictum that "philosophy of science without history of science is empty; history of science without philosophy of science is blind" (Lakatos, 1971, p. 102). We think a similar dictum applies to the issue we are addressing here: Scientonomy without scientometric analysis is empty; scientometric analysis without scientonomy is blind. Scientonomy aims to deliver a general descriptive theory of scientific change. But it is important to note that scientific publication practices in the 21st century are operating with unprecedented speed. It is thus a very challenging task to track all the relevant scientific articles and analyze their patterns of change without some advanced help from contemporary informational technology. This is why scientonomy without scientometrics is empty. If scientonomists aim to analyze patterns of change in contemporary science, they need the results of the large-scale analyses of scientometrics. Our HRV case is an example. Without the qualitative and qualitative results from Section 2, scientonomists would not have known where to apply the third law to the HRV case, not to mention where to empirically test the third law.

On the other hand, scientometrics is the field aspiring to employ statistical methods to measure the impact of scientific articles and understand the significance of scientific citations. But scientometrists need interpretive resources to help them interpret the quantitative results of statistical analyses. Tables 1 and 2, are an example. Though scientometrists can apply their tools to deliver the results in Tables 1 and 2, they need conceptual resources to help them generate a hypothesis concerning the impact of RA-11 and interpret the significance of citing RA-11 in the methods section of other papers. This is why scientometric analysis without scientonomy is blind because scientonomy provides the kind of conceptual resources scientometrists need to interpret their qualitative results.

In conclusion, we have illustrated how to integrate scientometrics with scientonomy with our HRV case. We have shown how scientonomists can use the scientometric results in Section 2 to identify where to apply the third law within a large body of HRV literature. We have also shown how the application of the third law to the HRV literature generates an empirical hypothesis that is further testable by scientometric tools. We think an integrated approach to the study of scientific change in a contemporary academic context will be fruitful.

Bibliography

Agosti, M., Borbinha, J., Kapidakis, S., Papatheodorou, C., & Tsakonas, G. (Eds.) (2009). *Research and Advanced Technology for Digital Libraries.* Springer.

Barseghyan, H. (2015). *The Laws of Scientific Change.* Springer.

Bechtel, W. & Richardson, R. C. (1993). *Discovering Complexity: Decomposition and Localization as Strategies in Scientific Research.* MIT Press.

Bigger, J. T., Fleiss, J. L., Steinman, R. C., Rolnitzky, L. M., Kleiger, R. E., & Rottman, J. N. (1992). Frequency Domain Measures of Heart Period Variability and Mortality after Myocardial Infarction. *Circulation,* 85(1), 164-171.

Caneva, K. (2012). What in Truth Divides Historians and Philosophers of Science? In Mauskopf & Schamltz (Eds.) (2011), 49-57.

Cartwright, N. (1983). *How the Laws of Physics Lie.* Clarendon.

Cartwright, N. (1999). *The Dappled World: A Study of the Boundaries of Science.* Cambridge University Press.

Chang, H. (2011). The Philosophical Grammar of Scientific Practice. *International Studies in the Philosophy of Science,* 25(3), 205-221.

Cicone, A. & Wu, H.-T. (2017). How Nonlinear-Type Time-Frequency Analysis Can Help in Sensing Instantaneous Heart Rate and Instantaneous Respiratory Rate from Photoplethysmography in a Reliable Way. *Frontiers in Physiology,* 8.

Djulbegovic, B. & Guyatt, G. (2019). Evidence vs Consensus in Clinical Practice Guidelines. *JAMA,* 322(8), 725-726.

Dupré, J. (1993). *The Disorder of Things: Metaphysical Foundations of the Disunity of Science.* Harvard University Press.

Golinski, J. (2012). Thomas Kuhn and Interdisciplinary Conversation: Why Historians and Philosophers of Science Stopped Talking to One Another. In Mauskopf & Schamltz (Eds.) (2011), 13-28.

Herring, E., Jones, K. M., Kiprijanov, K. S., & Sellers, L. M. (Eds.) (2019). *The Past, Present, and Future of Integrated History and Philosophy of Science.* Routledge.

Hon, E. H. & Lee, S. T. (1963). Electronic Evaluations of the Fetal Heart Rate. VIII. Patterns Preceding Fetal Death, Further Observations. *American Journal of Obstetrics and Gynecology,* 87(November), 814-826.

Honnibal, M. & Montani, I. (2017). *SpaCy 2: Natural Language Understanding with Bloom Embeddings, Convolutional Neural Networks and Incremental Parsing.* Retrieved from: https://spacy.io/.

Keyl, C., Lemberger, P., Pfeifer, M., Hochmuth, K., & Geisler, P. (1997). Heart Rate Variability in Patients with Daytime Sleepiness Suspected of Having Sleep Apnoea Syndrome: A Receiver-Operating Characteristic Analysis. *Clinical Science (London, England: 1979),* 92(4), 335-343.

Kleiger, R. E., Miller, J. P., Bigger, J. T., & Moss, A. J. (1987). Decreased Heart Rate Variability and Its Association with Increased Mortality after Acute Myocardial Infarction. *The American Journal of Cardiology*, 59(4), 256-262.

Kuhn, T. S. (1977). *The Essential Tension: Selected Studies in Scientific Tradition and Change*. University of Chicago Press.

Lakatos, I. (1971). History of Science and its Rational Reconstructions. In Lakatos (1978), 102-138.

Lakatos, I. (1978). *Philosophical Papers, Volume I*. Cambridge University Press.

Leydesdorff, L. & Milojević, S. (2013). Scientometrics. *ArXiv*. Retrieved from: http://arxiv.org/abs/1208.4566.

Lopez, P. (2009). GROBID: Combining Automatic Bibliographic Data Recognition and Term Extraction for Scholarship Publications. In Agosti et al. (Eds.) (2009), 473-474.

Malik, M., Farrell, T., Cripps, T., & Camm, A. J. (1989). Heart Rate Variability in Relation to Prognosis after Myocardial Infarction: Selection of Optimal Processing Techniques. *European Heart Journal*, 10(12), 1060-1074.

Massimi, M. (2018). Four Kinds of Perspectival Truth. *Philosophy and Phenomenological Research*, 96(2), 342-359.

Mauskopf, S. & Schamltz, T. (Eds.) (2011). *Integrating History and Philosophy of Science: Problems and Prospects*. Springer.

Miller, D. M. (2012). The History and Philosophy of Science History. In Mauskopf & Schamltz (Eds.) (2011), 29-48.

Mitchell, S. D. (2009). *Unsimple Truths: Science, Complexity, and Policy*. University of Chicago Press.

Nicolini, P., Ciulla, M. M., Malfatto, G., Abbate, C., Mari, D., Rossi, P. D., Pettenuzzo, E., Magrini, F., Consonni, D., & Lombardi, F. (2014). Autonomic Dysfunction in Mild Cognitive Impairment: Evidence from Power Spectral Analysis of Heart Rate Variability in a Cross-Sectional Case-Control Study. *PloS One*, 9(5), e96656.

Raimundo, R. D., de Abreu, L. C., Adami, F., Marques Vanderlei, F., Dias de Carvalho, T., Lessa Moreno, I., Xavier Pereira, V., Engracia Valenti, V., & Akemi Sato, M. (2013). Heart Rate Variability in Stroke Patients Submitted to an Acute Bout of Aerobic Exercise. *Translational Stroke Research*, 4(5), 488-499.

Rajendra Acharya, U., Paul Joseph, K., Kannathal, N., Lim, C. M., & Suri, J. S. (2006). Heart Rate Variability: A Review. *Medical & Biological Engineering & Computing*, 44(12), 1031-1051.

Rupik, G. (2019). Scientonomy: A Bold New Vision for an Integrated History and Philosophy of Science. In Herring et al. (Eds.) (2019), 19-37.

Task Force of the European Society of Cardiology and the North American Society of Pacing and Electrophysiology. (1996). Heart Rate Variability: Standards of Measurement, Physiological Interpretation, and Clinical Use. *European Heart Journal*, 17(3), 354-381.

Weissman, A., Torkhov, O., Weissman, A. I., & Drugan, A. (2009). The Effects of Meperidine and Epidural Analgesia in Labor on Maternal Heart Rate Variability. *International Journal of Obstetric Anesthesia*, 18(2), 118-124.

Wimsatt, W. C. (2007). *Re-Engineering Philosophy for Limited Beings: Piecewise Approximations to Reality*. Harvard University Press.

Yan, K., Tsai, M.-L., & Huang, T.-R. (2020). Improving the Quality of Case-Based Research in the Philosophy of Contemporary Sciences. *Synthese*, April. Retrieved from: https://link.springer.com/article/10.1007/s11229-020-02657-5.

Zammito, J. H. (2004). *A Nice Derangement of Epistemes: Post-Positivism in the Study of Science from Quine to Latour*. University of Chicago Press.

Chapter 5

Reconceiving Scientific Mosaics: A New Formalization for Theoretical Scientonomy

William Rawleigh

University of Toronto

Abstract: A central concept in scientonomy is the *scientific mosaic*, a concept intended to capture the state of an agent's scientific knowledge at a given point in time. The currently accepted definition of the *mosaic*, "a set of all epistemic elements accepted and/or employed by an epistemic agent", is a syntactic definition that fails to provide a theoretically robust investigational framework that allows scientonomists to explore mosaics and their minutiae fully and fruitfully. Pressingly, this definition leaves open some troubling semantic questions about the nature of *deducibility* and *meaning* within scientonomy's theoretical framework. This paper tackles these problems by proposing a semantic foundation for theoretical scientonomy rooted in explicitly set-theoretic concepts. It begins by examining a problem of semantic recursion through self-reference posed by the current definition and shows how the syntactic definition is fundamentally unable to overcome this problem. Instead, it proposes a semantic definition of mosaics which, while recursive, is not viciously or self-referentially recursive. It argues that by formalizing mosaics semantically as natural language models for scientific communities, scientonomy can overcome its semantic problems while also illuminating how truth-value assignment, inference, and the operation of higher-order laws work within theoretical scientonomy. To that end, it shows how a semantic definition can solve an outstanding problem with the third law having to do with the concept of deducibility.

Keywords: scientific mosaic; model theory; semantics; structure of scientific theories; derivability

***

1. Introduction

Constructing an adequate theoretical framework for formally representing scientific theories has been an enigmatic enterprise for philosophers of science since at least the logical positivists of the early twentieth century (Carnap, 1968; Pincock, 2009). Attempts to formally characterize the logical structure of organized scientific knowledge can even be dated as far back as Aristotle (Beth, 1959). The so-called "received view" of scientific theories – more generally referred to as the syntactic view and advocated for by the logical empiricists – prevailed until falling under sharp scrutiny in the 1950s and 60s, eventually falling out of favor along with the logical empiricist program more generally (Feigl, 1970).

Scientonomy answers the more general question of how to represent scientific theories and worldviews using the concept of a *scientific mosaic*. Defined at present as "A set of all epistemic elements accepted and/or employed by an epistemic agent",[1] the concept of the scientific mosaic is meant to capture the complete state of a scientific worldview in a single term. At any given time and for any given community the mosaic includes all of that community's beliefs about the natural world. While this rough definitive framework of the mosaic has been sufficient so far, scientonomy has so far failed to formalize the concept of a mosaic in a way that can fruitfully analyze the implicit logical structures inherent to the concept.

The formal structure of scientific mosaics – and the pragmatics associated therewith – have left some open problems that at present we lack the tools to solve. The accepted definition of a mosaic leaves some formal/theoretical problems in the theory of scientific change that need a solution. For one, the current definition of a mosaic along with the present formulation of the third law[2] leave us without a proper understanding of how *deducibility* works within the scientific mosaic. It also leaves us with no understanding of the rules of inference that must operate within the mosaic and leaves the exact logical and semantic nature of key elements undefined.

In this paper, I will propose a novel formalization of scientific mosaics that will give theoretical scientonomists new tools with which to think of the logical and semantic structure of epistemic elements. I will explain how adopting an explicitly set-theoretic notion of the mosaic that treats mosaics as natural language models for scientific communities allows us to overcome some core

---

[1] https://www.scientowiki.com/Scientific_Mosaic
[2] https://www.scientowiki.com/The_Third_Law_(Sebastien-2016)

theoretical issues that plague scientonomy as presently construed while allowing us to maintain the advantages that the accepted epistemic elements of our ontology afford our theoretical system. This conception will help to reveal the semantic workings of the mosaic, including how truth-values are assigned to sentences and sets of sentences, how inferences are drawn within the frameworks of particular first-order scientific theories, and how the second-order laws of scientific change operate on the mosaic. I will conclude by offering some examples of how this framework can help to resolve some of the outstanding issues in scientonomy by examining the third law.

2. The Logic of the Mosaic and a Recursion Problem

The *scientific mosaic* is the organizing unit in scientonomy and is broadly understood as the organizing object that undergoes scientific change. The presently accepted definition is from Barseghyan (2018), where a mosaic is defined as "a set of all epistemic elements accepted and/or employed by the epistemic agent".[3] In set-theoretic terms, we can think of the mosaic as being the unity of the primary epistemic elements of the ontology of scientific change. This conception of scientific mosaics has provided the guiding understanding of how scientonomy represents the process of scientific change. Discrete elements are understood as entering and leaving the mosaic according to the laws of scientific change, specifically the second law governing theory acceptance and the third law governing method employment.

There are a few important things to note about the presently accepted definition in terms of its content and structure - the first being that it is an explicitly set-theoretic concept. Given our current definition we can represent a mosaic in the language of set theory as follows:

$$\{M|e\} = \{A(t)e\} \cup \{A(q)e\}$$

where M is the mosaic borne by the epistemic agent, e is the epistemic agent in question, $A(yx)$ is a two-place predicate signifying that x accepts y, t is the set of theories, and q is the set of questions. The second thing to note is that the mosaic is a syntactic structure with no semantic content, inasmuch as t and q are just placeholders for whatever theories are accepted and methods employed a posteriori. Which questions and which theories are accepted is contingent on time, place, and community. They are not contained within the notion of the mosaic itself but come from without. The last thing to note about a mosaic is that it is agnostic as to what relation it bears with the external world

---

[3] https://www.scientowiki.com/Modification:Sciento-2018-0009

– who in particular bears mosaics, how, when, and under what conditions are all outside the concept.

This is a conception of a mosaic that is quite minimalistic. It stipulates very little and leaves much work to be done outside of itself – all the concept of a mosaic says is that it is a set-theoretic construct composed of employed methods and accepted theories. In particular, this set-theoretic, syntactic, and minimal notion of a mosaic lacks any kind of formal semantics for defining which propositions are true and which are false. So far theoretical scientonomy has treated this as being entirely determined by scientific communities themselves; whether a proposition is true or false depends on whether the community accepts the theory of which the proposition is a part. Consider a simple English sentence and its first-order predicate symbolization:

It is falling down

$\exists x(Hx \ \& \ Mx)$

Without a semantic interpretation, these sentences lack any meaning or truth value. In the practice and pragmatics of science, we depend on our scientific theories to play the role of assigning an interpretation to simple first-order sentences and the specific interpretation that we employ will be dependent on the theory that we accept.[4] If we assume that we accept something like an Aristotelian physics then the interpretive function will assign the following values: x specifies a member of the class of objects, H is a unary predicate indicating that x is composed primarily of heavy matter and M is a unary predicate indicating that x is in a state of natural motion. Read in plain English, the sentence says that there is an object that is composed of heavy matter in natural motion, which is consistent with the Aristotelian explanation for falling objects. Likewise, we can assign a Newtonian interpretation with different values: x specifies a member of the class of massive objects, H is a unary predicate indicating that x is being acted on by the gravitational force, and M is a unary predicate indicating that x is in motion. Again, a natural language interpretation says that there exists an object that is moving due to the gravitational force.

There are many things that have gone unmentioned in the semantic interpretation that I have sketched so far; in particular, I have made no

---

[4] See Hintikka (1998) in which he shows how Ramsey sentences cannot eliminate theoretical structures. It follows from this that if theoretical structures cannot be eliminated from the syntax, then they must be mapped semantically if sentences are to have any extra-logical meaning.

mention of how truth values are assigned, how truth-value operations are meant to work, or how truth valuations change. It is likewise mute about the role of indexes and how we might specify a given index. All I have shown is that the simple concept of the mosaic is semantically empty and that if we want to make sense of sentences, then we must refer to the *contents* of the mosaic, i.e., the theories of which the mosaic is composed. The tempting thing to do here is to simply refer to the external world to fill in the semantic gaps. What I mean by this is that the temping move is to simply say that in the case of sentences like "*It is falling down*" the semantic interpretation is determined by the referents. In this case, the sentence is true just in case the referent is in fact falling down. Similarly, sentences expressing something theoretical rather than observational. A sentence like "*electrons have a charge of negative one*" will have their semantic interpretation determined by their referent and will be true just in case electrons actually have a charge of negative one. Unfortunately, the historical-linguistic fact that theories have no stable referents over time, combined with the epistemic fact of scientific fallibilism forbids us from taking this escape route. Thus, we have no choice but to rely on the theories themselves when determining the semantic content of sentences.

The reliance on theories for semantic content in natural language interpretation is troubling, however. The presently accepted definition of theories comes from Sebastian and defines theories as sets of propositions (Sebastien, 2016). This definition is a markedly syntactic conception of theories. It says nothing about the interpretations, indexes, or semantics of the members of the sets in question. In fact, it tells us nothing at all except that they are propositional in nature. We could conceivably recast theories in the language of first-order predicate logic with a simple interpretation function but that would tell us very little except to define the variables and predicates without telling us anything about the truth-functional relationships between the variables, predicates, and connectives that are employed.

What is even more concerning is that the definition of the mosaic and the definition of a theory seem to leave us with a problem: the semantic content of theories is determined recursively by the theories themselves (Figure 5.1).

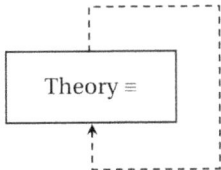

Figure 5.1: The semantic content of theories is determined recursively

Notice how given our current definitions the mosaic is composed of theories, but when we make natural-language interpretations on sentences in the mosaic we must rely on those same theories for our interpretation function. Consider another natural language sentence – this time one that expresses something theoretical rather than observational (since our accepted definition of theories is explicitly set-theoretic, it makes no difference what *kind* of sentence we use) – and its first-order predicate logic equivalent:

Massive objects curve spacetime

$\forall x \forall y (Gx \rightarrow Cy)$

This is immediately recognizable as one of the core theses of general relativity theory, which is part of the contemporary scientific mosaic. It is one proposition out of many propositions that jointly constitute the complete theory of general relativity, so if we let the sentence *'massive objects curve spacetime'* be given by α and general relativity be given by β, then α ∈ β, and β is a subset of the (contemporary) mosaic, given by M. The problem is that we already know that sentences in M, as syntactic structures, rely on the theories for their interpretive functions. In order for the sentence $\forall x \forall y (Gx \rightarrow Cy)$ to have any meaning at all we need to be familiar with general relativity (which assigns x to be a variable selected from the class of objects, y to be a variable selected from the set of spacetime coordinates, G to a unary predicate indicating massiveness, and C to a unary predicate indicating curvature), we need the theory to already have sufficient semantic richness to assign an interpretive function to α. Thus, the present definitions of a theory and of the mosaic jointly entail a recursion problem that indicates the need for a non-recursive semantic solution.

3. The Structure of Theories in Scientonomy

In order to develop a non-recursive solution to the mosaic problem we should be clear on how scientonomy construes the structure of theories, at least if we are to base the semantic interpretation of sentences on our accepted theories, as it seems natural to do. If we just go on the definition of a theory from Sebastian, then we have as our starting point that theories are (non-empty) sets of propositions (Sebastien, 2016). As noted above, this is a markedly syntactic conception. That we have this view of theories is a potential source of criticism for the whole analytic framework. It is suspiciously familiar to the logical positivist program that sought to recast scientific theories in terms of Ramsey sentences; sentences in first-order predicate logic that served to reconstruct scientific theories as being built up from observation sentences. While our chosen framework is free of the empiricist baggage that came with the positivist project, it seems that we nevertheless must deal with the same criticisms that have been lodged against the syntactic view of theories. Winther (2016) has an excellent summary of the problems facing the syntactic view, but suffice to say that the broad consensus among philosophers of science is that the so-called *received view*, as it was dubbed by Quine, fails to accurately represent the structure of scientific theories (da Costa & French, 1990; Lutz, 2014). Of these, two seem particularly troubling for scientonomy: first, that the syntactic view, such as it is, has no clear way to individuate theories that are syntactically identical but semantically distinct. Routley in his (1989) gives several examples of multiple intentionalities in logic and its consequences (Routley, 1989). Perhaps more seriously, the syntactic view has been charged – rightly – with ignoring the practice and history of science or, as Craver calls it, with ignoring the "structure of theories in the wild" (Craver, 2002).

The recognition of these issues has given rise to an alternative conception of theories in the literature: a semantic view of theories based on Tarski's satisfaction definition of truth (García-Carpintero, 1996). While there have been many variations of the semantic view from philosophers such as Suppes, van Fraassen, and da Costa and French, I will focus on that variation which has arguably been the most influential in the field of HPS – Giere's representational conception. First introduced in Giere's landmark paper "How Models are Used to Represent Reality" (Giere, 2004), the broad strokes' is that there is a two-place relationship between the world and the language scientists use to describe the world which Giere calls "representing". According to Giere, the key relationship is an ordered quadruple of the following form (Giere, 2004, p. 743):

S uses X to represent W for purpose P

In the above expression S is said to be the agent doing the representing – scientists – while X is the linguistic entity being used, W is the world, and P is the reason. To represent is to pair-up the linguistic entities with their counterparts in the world, so at an intuitive level, we can say things like "sociologists use the graph to represent changes in population for visualization purposes" or "biologists use the atomic formula to represent the structure of DNA for understanding transcription". Models here do the work of pairing linguistic structures with the theoretical entities they represent, i.e., the work of showing how S uses X. Thus, under Giere's framework, scientific theories are structured as abstract, non-linguistic (Craver, 2002, p. 58) semantic entities that serve as representational intermediaries.

The great insight from Giere is in the role he gives theories in mediating the relationship between the speakers of a language and the world. The role of a theory – as a linguistic entity – in scientonomy is to signify the beliefs of a community about the natural world, so there is an obvious symmetry in how Giere deploys the notion of a scientific theory and how scientonomy does. Of course, there is also a significant difference: Giere can depend on the pragmatic structures to supply theories with semantic interpretations. The great challenge that Giere faces with his conception is coming up with a formalization for the pragmatic elements of his quadruple, namely the W and the P. Formalizing and describing these elements lead Giere to adopt a more pragmatic stance towards the parts of the world being represented and the reasons of the actors for doing so, although it also enabled him to avoid the major problem with recursion that we are presently faced with. Since the theory of scientific change does not presently incorporate any of these pragmatic elements, we cannot solve the problem in the same way as Giere.

Fortunately for us, we are not quite looking at the same problems. Giere wants to see how scientists, from their own point of view, use models to represent reality. Our project is to represent how scientists represent reality, and by positioning ourselves at a level of analysis above the level at which science takes place and utilizing a nomological framework that gives us a clear meta-structure for examining theories, we can take the step that Giere could not without needing to totally revise how we understand theories in the first place. Taking our level of analysis one level up has the significant benefit of letting us keep our theories as they are without getting stuck in the metaphorical mud of pragmatics. Which exact parts of the world are being represented, why they are being represented as they are, the purpose for which they are being represented, and the modes of representation themselves are all immaterial to us as they fall outside the scope of a general descriptive theory of scientific change (see Barseghyan, 2015, specifically the section on scope). Thus, the definition of theories – as sets of propositions – can remain as it is so long as we find a different entity that can supply the mosaic with its semantic content that is not the theories themselves. This enables us to overcome the recursion problem.

4. A Non-(Viciously)-Recursive Semantics for the Mosaic

For the semantics of theories to be determined non-recursively, we can consider an alternative way of looking at the mosaic: rather than seeing the mosaic as a simple set-theoretic unity of elements, we can consider the mosaic to be a model for interpreting all-natural language sentences, whether those be observational, theoretical, or simply ordinary conversational sentences. So conceived, we can define the mosaic as being *a model of all accepted and employed epistemic elements* (Figure 5.2).

Scientific Mosaic ≡
A model of all epistemic elements accepted or employed by the epistemic agent.

Figure 5.2: A model-theoretic definition of 'scientific mosaic'

The major difference between this definition and the current definition is that it moves explicitly from the set-theoretic wording "set of all epistemic elements" to a semantic "model of all accepted elements". The broad definition of *theory* does a lot of work for us here; where a narrower definition of theory would pose a problem because we would *only* be modelling theories and questions, the definition of theories as simply "sets of propositions" lets us effectively treat any non-inquisitive sentence as a theory for our purposes. Accepting a theory, with this definition, simply means to regard it as true. Even more importantly, it lets us stipulate a language L such that L is the set of all theories. In other words, L is the entire language and L has very few limitations. Other than needing to be countable, L can include any variable, n-ary predicate, function, or connective that is actually used, and we need not worry about getting into the weeds of differentiation. So, we can begin by saying that M is a model of L, which gives us a starting point from which we can stipulate other features of what we can call the model-mosaic (MM).

In the language of set theory, this definition lets us say that MM is the *structure* of the set of all epistemic elements. The epistemic elements – questions and theories – are themselves the *signature*; we can denote the signature with ε. When we use ε, what we are really using is a shorthand for the set of all the constants, variables, predicates, functions, and combinations thereof that are members of each singular epistemic element, be it a question or a theory. This is allowed by syntactic nature of the epistemic elements. If each epistemic element is given by φ, then the constants c, variables v, predicates P, and functions F will all be members of φ. This tells us that φ will be a member of ε, and that MM is a structure of ε. Now if we take any proposition ψ, we can say that MM models ψ just in case ψ is true for some community C (the index) given the assignments of c, v, P, and F under ε.

This is all highly abstract, so we can ground it in a familiar example. Suppose we let the sentence '*massive objects curve spacetime*', which was used in an earlier example, stand in for ψ. Syntactically, this is equivalent to:

$$\forall x \forall y (Gx \rightarrow Cy) = \psi$$

The variable terms are 'objects' and 'spacetime', the predicate terms are 'massive' and 'curves', and the constant is the material conditional. The sentence does not contain any functions. Suppose also that the community to which we are indexing is the contemporary physics community, and the mosaic contains all the propositions of general relativity. Since *massive objects curve spacetime* is itself one of the propositions of general relativity, the sentence is a member of the signature and so is trivially modelled by the mosaic for the community. If, however, the community to which we are indexing is the Newtonian community in England circa 1720 CE, we come up with a different result. In this case, the sentence is not trivially true by virtue of being part of the signature because the Newtonian community had a different mosaic. Thus, when we assign the meanings of 'object' and 'spacetime' to the variables and 'massive' and 'curve', the interpretation is now false given our indexed community, and MM does not model the sentence. This allows us to say that the sentence is true *because* it stands in a certain sematic relation between and with respect to the structure (the contents of the mosaic) and the index (the community in question). All of these pieces are indispensable when it comes to determining the actual truth-values of a given proposition or theory when we speak of it in a scientonomic context.

We can also do this with observation terms. Take the sentence '*the light bent around the star*'. Let this sentence be given by

$$\exists x \exists y (Sx \,\&\, \gamma y \,\&\, Byx) = \psi$$

This is not one of the core propositions of general relativity and cannot be trivially true, but the sentence turns out to be true as long as the variable assignment, predicate assignment, operators, and constants of the structure (the mosaic) make the sentence turn out true given the index (the epistemic community in question). We can use the expression

$$MM \vDash_C \psi$$

to express that MM models ψ for the community in question, which is to say that ψ turns out to be true. When MM is a mosaic that includes the general theory of relativity and orthodox quantum mechanics and C is the international physics community circa 1930 CE, the sentence comes out true. The signature of MM picks out the class x as the class of large objects in space and y as the class of gauge bosons. It assigns S as a single-placed predicate meaning 'is a star', γ as a single-placed predicate meaning 'is light', and B as a two-place predicate meaning that 'y bends around x'. It also includes the conjunction as a constant. When this signature is taken together with the indexed community the sentence comes out as true.

What is noteworthy in the above cases is that in order for the sentence to come out true it is not independently sufficient that the structure assigns certain variables, or that the indexed community is a certain epistemic community (the inter-war western physics community, for example). Rather, both the structure and the community must be correct. If either the structure or the community changes, then the sentence turns out to be false. This seems trivial – after all, it only makes sense that changing the structure to Aristotelian mechanics or the community to English Newtonians would cause a sentence like *'the light bent around the star'* to come out false. However, it is significant if we do not want there to be a conflict between the historical record and the semantics that underlie scientonomic theory. If it *were not* the case that changing either the structure or the index changed the truth-values of the sentences, then we would be in major trouble. Continuing with the example of *'the light bent around the star'*, we want intuitively to say that if we changed the contents of the structure so that the mosaic no longer included variable and predicate assignments from general relativity but instead included variable and predicate assignments from say, Newtonian mechanics, that the sentence would turn out false. Obviously, no sentence in Newtonian mechanics that includes the idea of light bending should turn out to be true, since no Newtonian would accept any such proposition, and the semantics of our theory should work out so this is the case. Likewise, we do not want it to happen that if the index changed that the truth-value assignments would come out the same. If we change the index so that we are indexed to the community of Cartesian natural philosophers circa 1680, we do not want it to be acceptable that light would bend around stars, even if the variable assignment from the mosaic makes it so. This is because the community of Cartesian natural philosophers *would not have made this observation*. Indeed, they *could not* have made this observation! Consequently, a given sentence can only be true if both the community and the mosaic dictate that it be so.

This way of understanding the mosaic is still recursive to an extent, but the way that it is recursive is not problematic. This is because, unlike the kind of recursion problem posed by a strictly set-theoretic notion of the mosaic, the recursiveness present in the model-theoretic understanding is not self-supporting. The kind of recursion present in the model-theoretic understanding of the mosaic is a kind of two-place recursion that has been widely accepted in epistemology and the philosophy of science in decades past. To see why, consider Figure 5.3 that represents both cases. As long as we avoid the situation in the diagram on the left and have only the kind of recursion pictured in the diagram on the right, we have no problems. The same kind of recursion that I am proposing here was relied upon by Ernest Sosa in his presentations of formal foundationalism, coherentism, and reliablism (Sosa, 1980), and was presented by Yehoshua Bar-Hillel in the early 1950s as a mode of understanding definitions in the empirical sciences (Bar-Hillel, 1953; 1954). Instead of theories depending on themselves for semantic content, theories depend on the mosaic as an interpretive function and mosaics in turn depend on theories to guide interpretations (see above diagram).

Figure 5.3: Two types of recursion

Notably, there is no need for the mosaic to be sound as there is no reason why we should expect the mosaic to forbid the entailment of contradictory sentences a priori. Likewise, we have no reason to expect that the mosaic would necessarily be complete, since such an expectation would seem to be fundamentally at odds with the well-entrenched principle of fallibilism in empirical science.[5] Certain very strict interpretations on certain languages may forbid the derivation of contradictory sentences, but the existence of such cases need not necessarily concern us. To see why we need only look at the history of science and the fact that the same mosaic has, at various times, included theories which entail contradictory sentences. Arguably, the most famous case where two theories imply contradictory conclusions is the case of general relativity and quantum mechanics, both of which make contradictory predictions in very strong gravitational fields in very small regions of spacetime, such as those around the event horizon of a black hole (Plato, Hughes, & Kim, 2016). Historians and philosophers of science have documented scores of such cases, where the theories accepted by various scientific communities have been inconsistent (Batterman, 2014). The apparent paraconsistency in both syntax and semantics of actual mosaics a posteriori suggests that any call for either soundness or completeness a priori neglects the behavior of the actual agents the system attempts to model. That some agents are tolerant of inconsistent theories and others are not is simply a

---

[5] While it is tempting to raise the specter of Gödel when discussing completeness, the fact is that no scientific mosaic *necessarily* requires either consistent axioms or that some amount of arithmetic be derivable. The fact that we never observe actual mosaics that satisfy the requirements of completeness, as well as the fact that nothing seems to demand that mosaics meet those conditions is sufficient.

feature of a given agent and says nothing about the inherent structure of all mosaics. Sarwar and Fraser have treated the subject at length, so further explication is not necessary here (Fraser & Sarwar, 2018). What remains is to see what this new conception of the mosaic can contribute to scientonomy.

5. Some Consequences of a Model-Semantics for Mosaics

That the semantics of the mosaic is explicitly model-theoretic would have several interesting implications, specifically for how we understand the third law. At present, there are several problems with how the third law is formulated and operates that need to be resolved. Foremost among these is that the third law is, in its current form, based on an outdated ontology that assumes that *methods* of theory evaluation are a fundamental member of our ontology of epistemic elements. After the acceptance of Barseghyan's proposal (Barseghyan, 2018) that methods be subsumed under the category of normative theories (a consequence of which is that methods are a subset of *norms*), the third law no longer exhaustively covers any situation in which we *employ* any kind of normative theory. In its present form, it is limited to methods, though there is no strictly logical justification at present for thinking that the mechanism by which a method is employed is any different than the mechanism by which any other norm is employed.[6] Consequently, it seems that the first thing that we need is a formulation of the third law that acknowledges the employment of norms more broadly, rather than one that limits itself to methods. The easiest way to make this change is to simply substitute the word "method" in our present employment of the third law for the word "norm" (Figure 5.4).

3rd Law: Norm Employment
A norm becomes employed only if it is deducible from some subset of other employed methods and accepted theories of the time.

Figure 5.4: The first reformulation of the third law

---

[6] There is no "strictly logical" reason for thinking that epistemic norms, i.e. methods, are different from other kinds of norms. Semantically speaking any kind of norm – epistemic, political, moral, etc. – fulfils the same role in structuring the logical discourse, i.e. by specifying constraints that must be me met by an agent interacting with the system. Whether there is a difference *in practice* between epistemic norms and other kinds of norms is a more empirical matter.

While this formulation solves the problem of incongruity with our accepted ontology, it nevertheless leaves us with another significant problem in the meaning of the word *deducible*, which is one of the open problems currently accepted by our community. The trouble is that at present, the meaning of *deducible* and its conjugates with respect to the general theory of scientonomy is undefined; we do not currently know what it means for something to be deducible, what the criteria of deducibility would be, or whether the conditions of deducibility would be part of the first-order theories of the mosaic or part of the second-order theories that range over the mosaic. Answering these questions is one of the possible ways that we can apply a model-theoretic notion of the mosaic to the problems that we face in theoretical scientonomy.

We can begin to consider the problem of deducibility by straightforwardly applying the notion of deducibility that prevails in mathematical model theory. In that context, for some sentence to be deduced from some other sentence or set of sentences is a matter of semantic entailment. This means that in the expression

$$\Sigma_K \vDash \psi_K$$

we say that the set of sentences Σ entails ψ just in case whenever the truth valuation of every member of

$$\Sigma = (\varphi_1, \varphi_2, \dots \varphi_n)$$

is true, then ψ also has an assigned truth value of true on the structure K (recall that the structure is composed of the variable, constant, predicate, and operator assignments). In this context, there is no meaningful difference between saying that Σ entails ψ, or Σ implies ψ, or ψ is deducible from Σ since we do not draw a distinction between assertion, derivation, and other kinds of representation with regards to epistemic agents. At most, we might be able to draw a modal distinction between deducibility and entailment/implication since the former may be interpreted by using the modal operator for *possibility*, in that it is *possible* for an epistemic agent to deduce ψ from Σ, though they may not have done so. For our purposes, this distinction seems immaterial. All the outlined terms have an explicit connection with semantic entailment, and it is exactly the entailment relationship that is of concern to us. Drawing a further modal distinction may introduce a time index that we have an interest in dealing with, so for parsimony's sake, it seems best to regard them all as synonymous.

We can consider two practical examples of semantic entailment in science, one dealing with a descriptive theory and one dealing with a normative theory.

Beginning by looking at descriptive case, suppose first that we have a given set of sentences:

1) All particles with mass travel at sub-luminal speeds.

2) All protons have mass.

3) The particle accelerator fired a proton.

4) The proton fired by the particle accelerator travelled at subluminal speeds.

We can suppose that sentences (1)-(3) are the members of Σ, that sentence (4) is identical with ψ, and that the structure MM of the language L to which Σ belongs has assigned the variables, constants, operators, and predicates in such a way that Σ turns out to be true (we can suppose that the community is arbitrary and so is automatically congruent with MM). In this case, Σ ⊨ ψ is true just in case the rules governing the constants and operators in K guarantee the truth of ψ whenever Σ is true. If we suppose that the constants and operators of MM are the kinds of first-order predicate logic constants and operators that we are used to, then we can safely say that ψ is in fact semantically entailed by Σ. Of course, whether those operators and constants are in fact in use depends on the mosaic itself, since, as we established earlier, the epistemic elements of the mosaic are the signature that determines the permitted operators and constants on the structure MM. A different assignment of constants and operators may not guarantee the truth of ψ given the truth of Σ. As a consequence, it is clear that the model-theoretic conception of the mosaic makes the rules of inference part of the mosaic itself rather than second-order theories that range over the contents of the mosaic.

We can now turn our attention to a normative case. Suppose that we have the following set of sentences:

1) The world is composed of particles which can have an infinite number of arrangements.

2) Any observation can be accounted for after the fact by a certain arrangement of particles.

3) There is more to observable phenomena than can be directly observed with the senses.

4) A successful scientific theory should account for (1)-(3).

5) Requiring that new theories provide novel predictions that are confirmed through experiment adequately accounts for (1)-(3).

6)   All new theories should provide confirmed novel predictions.

Once again, we can suppose that sentences (1)-(5) are the members of Σ, that sentence (6) is identical with ψ, and that the structure MM of the language L to which Σ belongs has assigned the variables, constants, operators, and predicates in such a way that Σ turns out to be true. This case is distinct from the first case in that it seems to require a deontic operator that can properly account for the imperative nature of sentences (4) and (6) but is otherwise the same. So long as the signature consists of a term defining a deontic operator and rules of inference, the structure MM will have no trouble guaranteeing the truth of ψ based on the truth of Σ. As before, a different signature will leave us with a different result if it lacks a deontic operator or if the rules governing the deontic (or classical) operators in the sentences fail to guarantee the truth of ψ given the truth of Σ. Adding any kind of novel operator, be it modal, deontic, erotetic, or some other non-classical operator will pose challenges for formalization but will otherwise leave the underlying logic intact. Formalizing non-classical operators is strictly a practical problem for theoretical scientonomists and need not concern us as far as the semantic structure of the mosaic is concerned. As long as the structure can model the language, and there is no reason to suppose that any language exists which fails to be modellable, then our project remains undisturbed.

What the above discussion suggests is that in adopting a definition of the mosaic that accounts for its model-theoretic semantics, we should likewise adopt a definition of deducibility that accounts for the model-theoretic notion of semantic entailment. Since *derivable*, in the context of mathematical model theory, simply means *to be semantically entailed*, it seems that *derivable* and its conjugates are a better candidate term than *deducible*, if for no other reason than the term *deducible* has connotations that are associated with a strictly deductive logic, which may not be congruent with any particular mosaic. Even the mosaic of the contemporary scientific community seems to admit to inductive and abductive reasoning types, not to mention certain kinds of non-classical logics. Using the term *derivable* frees us from those connotations while also giving us a clear, in-context meaning that we adopt for our own ends. This means that we can define *derivable* in terms of our epistemic elements (Figure 5.5). In addition to clarifying that deducibility actually means derivability, this definition also offers the slight clarification that derivability strictly deals with derivation from a finite number of other elements. While hardly a practical concern, given the practical inapplicability of actual infinities, the stipulation that derivability deals only with finite cardinalities at least side-steps the theoretical hurdle that is raised by the specter of infinity.

Derivable ≡
An element of the mosaic is said to be derivable when it is a semantic consequence of a finite number of other elements of the mosaic.

Figure 5.5: A definition of 'derivable'

One thing to note is that once we have defined derivability thusly, a corollary arises based on how we came to that definition. Since the operators and constants – and by extension, the rules of inference – are part of the signature, which we defined as the epistemic elements of the mosaic, that means that the rules of inference governing the notion of semantic consequence are themselves part of the mosaic. That leads us to the *derivability corollary* (Figure 5.6). This corollary simply serves to clarify what we have already stated. Unlike the laws of scientific change themselves, which are second-order principles that describe the mechanics of the mosaic and operate at a level of analysis above the workings of science in the mosaic, the laws of inference that govern derivability are first-order theories that are part of the mosaic and can undergo scientific change according to the second law.

Derivability Corollary
The rules for determining whether an element is derivable from a set of other elements of the mosaic are themselves elements of the mosaic.

Figure 5.6: Derivability Corollary

Having clarified that *derivable* is a preferable term to *deducible*, we are now in a position to present a new formulation of the third law that is free from the defects of the present formulation. Recall that when we last discussed the third law at the top of this section, we settled on the idea that the term *norm* should be used in place of the term *method*, and we have now settled on the idea that the word *derivable* should take the place of *deducible*. This leads to a new formulation of the third law (Figure 5.7).

3rd Law: Norm Employment

A norm becomes employed only when
it is derivable from some subset of
other employed norms and accepted
theories of the time.

Figure 5.7: The second reformulation of the third law

While much better, of course, there is one final consideration; with the acceptance of *questions* into the epistemic elements of the ontology of scientific change, the elements of the mosaic are now more expansive than just theories and subtypes of theories (i.e., norms). As I have already noted, the signature of the mosaic could conceivably include a range of both logical and non-logical operators, which could potentially include erotetic operators. If erotetic operators were to be included in the signature then it would follow that there is a plausible situation in which norms could potentially be derived – at least in part – from *questions*, which means that a formulation of the third law that excludes questions would fail to comprehensively describe all cases of norm employment. Following Barseghyan (2018), we can solve this issue by simply replacing a specific enumeration of our epistemic elements with a general one. This would lead to another version of the third law (Figure 5.8).

3rd Law: Norm Employment

A norm becomes employed only if it is
derivable from a non-empty subset of
other elements of the mosaic.

Figure 5.8: The third reformulation of the third law

This version of the third law successfully incorporates everything that has been noted already, while maintaining the virtues laid out by Sebastian when she first proposed this revised formulation (Sebastien, 2016). It is consistent with the new ontology of epistemic elements, it is broad enough to be resilient to future changes in our ontology of epistemic elements, and its use of the term "non-empty subset" clarifies some outstanding problems relating to the third law. First, it decisively resolves the question of whether methods can be derived from empty sets by clarifying that no such thing is possible without there being a violation of the laws of scientific change. Second, it clarifies that it is at least *possible* for norms to be derived from any combination of types of epistemic elements. *In theory*, norms could be derived from any combination of descriptive theories, definitions, norms, or questions, including only one type or at least one of every type. While it is unlikely that we ever observe cases where norms are derived from at least a single kind of theory – most deontic logics seem to suggest that we need at least one descriptive theory and one norm – this formulation leaves open the possibility that a contrary situation might obtain. This is desirable because it prevents us from writing off potential cases of scientific change as unscientific on the sole basis of the rules of inference employed by an epistemic agent, a situation which is no less chauvinistic than writing off a case of scientific change because it is based on a theory we no longer accept. A final point in favor of this new formulation is that it expresses the derivability condition as a necessary condition for norm employment but not a sufficient one. It seems intuitively correct to think that there are cases in which a norm is derivable from some accepted elements and yet not accepted by the community. Whether such a case would qualify as an unscientific case of non-change in the mosaic, or simply be the result of an inconclusive evaluation is unclear. This formulation is agnostic about the existence or nature of particular sufficient conditions for norm employment and the topic may be a fruitful subject for future research.

6. Conclusion

The current syntactic conception of both mosaics and theories in scientonomy leaves much to be desired. On top of the many documented problems with syntactic views of theory choice, the present conception of the mosaic has a significant problem with recursive semantics that the current definitions and structures are ill-equipped to deal with. While these issues are undoubtedly significant, it must also be acknowledged that a great deal of our current body of scientonomic work relies on something like a syntactic conception of theories. By introducing an explicitly model-theoretic notion of scientific mosaics, I propose that we can resolve the issues presented by the syntactic nature of our mosaic while still being able to preserve the beneficial work that the syntactic notion does for our theory. Accepting this notion opens some fruitful avenues of research, which I have demonstrated by applying the model-theoretic conception of the mosaic to some existing problems surrounding the mechanism of norm employment. In doing so, I have proposed a novel definition of *derivability*, deduced from that a *derivability corollary*, and suggested a new formulation of the third law of scientific change. Embracing a more comprehensive and formal semantics for the mosaic allows scientonomists to not only be more rigorous in how we engage with scientonomy at both the theoretical and observation level and equips us with new tools to tackle some deep and intransigent problems facing our theory. I hope that by accepting these modifications the community will open up new doors for new research on the formal aspects of our epistemic elements.

Bibliography

Bar-Hillel, Y. (1953). On Recursive Definition in the Empirical Sciences. *Proceedings of the XIth International Congress of Philosophy*, 5, 160-165.

Bar-Hillel, Y. (1954). Logical Syntax and Semantics. *Language*, 30(2), 230-237.

Barseghyan, H. (2015). *The Laws of Scientific Change*. Springer.

Barseghyan, H. (2018). Redrafting the Ontology of Scientific Change. *Scientonomy*, 2, 13-38.

Batterman, R. W. (2014). The Inconsistency of Physics (with a capital "P"). *Synthese*, 191(13), 2973-2992.

Carnap, R. (1968). *The Logical Structure of the World*. Routledge & K. Paul.

Craver, C. F. (2002). Structures of Scientific Theories. In Machamer & Silberstein (Eds.) (2002), 55-79.

da Costa, N. C. & French, S. (1990). The Model-Theoretic Approach in the Philosophy of Science. *Philosophy of Science*, 57(2), 248-265.

Feigl, H. (1970). The "Orthodox" View of Theories: Remarks in Defense as well as Critique. *Minnesota Studies in the Philosophy of Science*, 4, 3-16.

Fraser, P. & Sarwar, A. (2018). A Compatibility Law and the Classification of Theory Change. *Scientonomy*, 2, 67-82.

García-Carpintero, M. (1996). What Is a Tarskian Definition of Truth? *Philosophical Studies: An International Journal for Philosophy in the Analytic Tradition*, 84(2, Analytic Philosophy in Europe), 113-144.

Giere, R. (2004). How Models are Used to Represent Reality. *Philosophy of Science*, 71(5), 742-752.

Hempel, C. G. (1989). Aspects of Scientific Explanation. In Kitcher & Salmon (Eds.) (1989), 331-489.

Hintikka, J. (1998). Ramsey Sentences and the Meaning of Quantifiers. *Philosophy of Science*, 65(2), 289-305.

Kitcher, P. & Salmon, W. (Eds.) (1989). *Scientific Explanation*. University of Minnesota Press.

Kuhn, T. (1970). *The Structure of Scientific Revolutions*. Chicago: The University of Chicago Press.

Lutz, S. (2014). What's Right with a Syntactic Approach to Theories and Models. *Erkenntnis*, 79, 1475-1492.

Machamer, P. & Silberstein, M. (Eds.) (2002). *The Blackwell Guide to the Philosophy of Science*. Blackwell Publishers.

McGrath, M. (2014). Propositions. In Zalta, E. N. (Ed.) (2014). *The Stanford Encyclopedia of Philosophy (Spring 2014 Edition)*. Retrieved from: https://plato.stanford.edu/archives/spr2014/entries/propositions/.

Norman, J. & Sylvan, R. (Eds.) (1989). *Directions in Relevant Logic*. Springer.

Oberheim, E. & Hoyningen-Huene, P. (2018). The Incommensurability of Scientific Theories. Zalta, E. N. (Ed.) (2018). *The Stanford Encyclopedia of Philosophy (Fall 2018 Edition)*. Retrieved from: https://plato.stanford.edu/archives/fall2018/entries/incommensurability/.

Pincock, C. (2009). Carnap's Logical Structure of the World. *Philosophy Compass*, 4(6), 951-961.

Pincock, C. (2015). Abstract Explanations in Science. *British Journal for the Philosophy of Science*, 66, 857-882.

Plato, A., Hughes, C., & Kim, M. (2016). Gravitational Effects in Quantum Mechanics. *Contemporary Physics*, 57(4), 477-495.

Rawleigh, W. (2018). On the Status of Questions in the Ontology of Scientific Change. *Scientonomy*, 2, 1-12.

Routley, R. (1989). Philosophical and Linguistic Inroads: Multiply Intensional Relevant Logics. In Norman & Sylvan (Eds.) (1989), 269-304.

Sebastien, Z. (2016). The Status of Normative Propositions in the Theory of Scientific Change. *Scientonomy*, 1, 1-9.

Shoenfield, J. R. (1967). *Mathematical Logic*. Addison-Wesley.

Sosa, E. (1980). The Raft and the Pyramid: Coherence versus Foundations in the Theory of Knowledge. *Midwest Studies in Philosophy*, 5(1), 3-26.

Winther, R. G. (2016). The Structure of Scientific Theories. In Zalta, E. N. (Ed.) (2016). *The Stanford Encyclopedia of Philosophy (Winter 2016 Edition)*. Retrieved from: https://plato.stanford.edu/archives/win2016/entries/structure-scientific-theories/.

Woodward, J. (2013). Mechanistic Explanation: Its Scope and Limits. *Proceedings of the Aristotelian Society*, Supplementary Volume, 39-65.

Woodward, J. (2017). Scientific Explanation. In Zalta, E. N. (Ed.) (2017). *The Stanford Encyclopedia of Philosophy (Spring 2017 Edition)*. Retrieved from: https://plato.stanford.edu/archives/spr2017/entries/scientific-explanation/.

Chapter 6

Episodic Rationality in Scientific Change

Patrick Fraser

University of Toronto

Abstract: One of the most salient lessons from HPS as a discipline is that science is a living, breathing endeavor; one whose rules and values are constantly changing. As such, there is an essential tension between the hope for a coherent, unified conception of scientific rationality on the one hand, and the recognition of the diversity of perspectives which fit into the framework called science. The big question, of which I hope to answer a small part, is: how can rationality and relativism be reconciled with one another? To do this, I present a rational reconstruction of a theory of scientific change which resembles Barseghyan's theory of scientific change. I interpret scientific knowledge modally; the scientific mosaic of a community at a particular time is taken to represent the actual instantiation of a collection of possible scientific changes, all linked to one another through a Kripkean semantics of possible worlds. I then draw a correspondence between accepted scientific theories and employed methods with logical axioms and rules of inference respectively and use this to construct a logical framework for studying the modality of scientific knowledge. I use this framework to obtain a notion of scientific rationality which is contextually localized, but still presents a clear direction of scientific development at every individual time step.

Keywords: theory change; acceptance; rationality; contingency; modal logic

***

1. Introduction

Like most issues in contemporary philosophy of science, understanding scientific change – the process by which scientific theories, methods, and values evolve – can be traced back to the logical empiricism of the early twentieth century. According to the logical empiricists, science was thought to be a rational, systematic manner of theorizing about the natural world. Thus, the ways in which science changes was thought to be a purely *logical* (i.e., analytic) process of probabilistic deductions with credence on our beliefs informed by experimental confirmation, which were taken to be purely observational (i.e., synthetic) (Carnap, 1937).

This idea of strictly logical, linear progression was challenged, for instance, by Quine (Quine, 1951) who made it clear that the analytic/synthetic distinction could not be upheld. Thus, science must be understood *holistically*, with explanations putting both theorizing about and observing the natural world on equal footing. A more elaborate and holistic theory of scientific change, due to Lakatos, overcame these difficulties (Lakatos, 1970). However, Lakatos's theory still treated science as a wholly abstract endeavor, quite detached from the human and social contexts in which it takes place.

In 1962, Kuhn famously made clear that the theoretical terms which our scientific theories rely so closely upon are indeed historically contingent (Kuhn, 1962). If we wish to compare two theories for the purposes of, for instance, assessing which to accept (which is the crucial 'rational' feature of scientific change held onto by the logical empiricists and Lakatos alike), we must first be able to translate between the theories in question. However, it is a historical fact, as Kuhn showed, that the terms of new theories often are not reducible to the terms of old theories; the two are *incommensurable* (Kuhn, 1982). Thus, if one wishes to give a rational account of why one theory was favored over another (e.g., why Lavoisier's chemistry won out over theories of phlogiston), they must do so using a full *historical* analysis of the two theories *in their own terms* and using the scientific method employed at the time. This last point, too, is importantly missing from most early accounts of scientific change; scientific methods themselves are prone to change too (Feyerabend, 1975). It has also been argued that scientific change is highly sensitive to the social *values* scientists place on themselves, which are of course also historically contingent (Merton, 1973).

In light of these complexities, Laudan provided the first notable theory of scientific change which took into account the changeability of scientific methods and the social influence of scientific values in his reticulated model (Laudan, 1984). This model then served as a starting point for Barseghyan's scientonomic approach to scientific change (Barseghyan, 2015).

The salient lesson from the scientonomic analysis of scientific change is that any adequate theory of scientific change must satisfy the following criteria. First, theory assessment must be carried out using the actually employed scientific methods of the appropriate community at the appropriate time. Additionally, these methods must not change arbitrarily, but in a manner which is logically coupled to the *previously* employed methods and accepted theories. Finally, it must be possible for scientific changes to be socioculturally influenced. However, if such a theory is to take the process of scientific change to be *rational* in some capacity, these sociocultural influences must be constrained in such a manner that they do not introduce inconsistencies; at a deeper level of analysis, they must still follow a collection of consistent logical rules. This last point shall be called the *rationality criterion* for theories of scientific change.

These considerations are broad. Scientonomy satisfies the first three of these requirements, however, it is not clear whether or not it satisfies the rationality criterion. In the following section, I provide a formal account of scientific theory change in terms of modal logic from which the salient features of scientonomy (insofar as it pertains to theory change) may be recovered. I then demonstrate that such a formalism *does* satisfy this rationality criterion. In particular, I show that it exhibits what I call *episodic rationality*; the meaning of the term 'rationality' may be prone to change in this context, but in each episode, it is well defined, and is found in scientific change. One may then interpret the below framework either as a rational reconstruction of a modified version of scientonomy, or they may instead view it as a starting point for a more fully-fledged multi-dimensional formal theory of scientific change which could feasibly recover scientonomy along one axis.

2. Formal Framework

To begin, there are essentially two distinct kinds of scientific change; theory change, and method change. That is, scientific change (at least as it is understood in the scientonomic setting) is the term used to describe instances when a particular (scientific) community changes their accepted theories (by accepting a new theory and possibly rejecting an old one in the process), or their employed methods. Scientonomy is then a descriptive theory of which instances of these changes are, in some minimal sense, rational and scientific.

In scientonomy, a snapshot of science for some particular community takes the form of a mosaic. Scientific change is then the process by which one mosaic undergoes a transition into another (perhaps with a change to the underlying community along the way). To this end, there is an obvious sense in which one mosaic is *accessible* to another; we may say that a mosaic M' is accessible to a mosaic M and write $M \leq M'$ just in the case when there is some permissible piecemeal scientific change (whatever 'permissible' here means) which takes M to M', supposing the communities before and after the transition are, in some as of yet undetermined sense, sufficiently similar to each other.

Another feature of the mosaic-oriented approach to scientific change is that the scope of possible mosaics extends far beyond actual, historically realized mosaics. After all, the process of scientific change is underdetermined; there are many permissible scientific changes which simply never occur. As such, it is clear that there is a built-in modality in every mosaic which is actually realized; each carries with it a broad range of accessible mosaics, only a small number of which will ever become actual (i.e., which will legitimately correspond to the scientific mosaic of an actual community).

Finally, mosaics carry with them an as of yet implicit logical structure. Theories within mosaics are treated as logical objects which, rather than being prescribed values of truth or falsity, are endowed with an acceptance value; any theory is either *accepted* relative to a mosaic or *unaccepted* relative to that same mosaic. Likewise, the manner in which these acceptance values are determined are given by the methods of that mosaic; methods serve as axioms and rules of inference, ensuring that scientific change is a rule-governed process.

What I have been hinting at in this discussion of mosaics is that they lend themselves quite naturally to a representation in terms of modal logic. The basic format of modal logic is the following.

Given a language of atomic symbols (i.e., p's and q's) and logical connectives $\vee$, $\wedge$, $\rightarrow$, and negation $\neg$, a structure of modal propositional logic (MPL) is a triple $\langle \mathcal{W}, \mathcal{R}, \mathcal{I} \rangle$ where $\mathcal{W}$ denotes a collection of *possible worlds*, $\mathcal{R}$ denotes a binary relation on $\mathcal{W}$ which reflects inter-world accessibility, and $\mathcal{I}$ denotes an interpretation function which allows truth to be defined for arbitrary logical formulas at arbitrary worlds. Such a system comes equipped with two additional logical operators, a possibility operator $\Diamond$ and a necessity operator $\Box := \neg \Diamond \neg$.

The semantics of modal propositional logic is such that, at a world w, given a formula ϕ, $\Diamond\phi$ is true exactly when there is some world u which is accessible to w such that ϕ is true in u. Necessity, then, is such that $\Box\phi$ is true in the world w only if ϕ is true in *all* worlds which are accessible to w. In what follows, I shall make use of an unquantified, predicated variant of usual MPL. See the Appendix for details of the syntax and semantics of this modal logic. See (Sider, 2010; Burgess, 2012) for an introduction to modal logic in general.

The way in which mosaics and scientific theories interact mirrors this structure very closely. Mosaics play the role of possible worlds, *possible* scientific change induces an accessibility between worlds, and acceptance valuation mimics truth valuation. Theories, therefore, play the role of formulas insofar as theories are assigned an acceptance value at specific mosaics just as formulas are assigned truth values at specific worlds. Hence, it is natural to seek a formal framework within which mosaics are directly represented as possible worlds and the full power of modal logic is then deployed to provide an account for scientific change under this correspondence. Such an account is the product of this article.

Two important qualifications are in order. First, in what is to follow, I only provide an account of *theory* change via this modal framework. That is, I represent all of the relevant features of scientific change thus mentioned in the language of modal propositional logic and proceed to provide an account for how any particular theory change may be understood robustly in these terms. I do not, however, provide an account of *method* change here; such changes turn out to occur at the *meta-logical* level of the framework provided here, and thus require additional tools to deal with appropriately (especially in light of the fact that these methods may be inductive in nature, and thus not easily accounted for in purely propositional or predicated deductive terms).

Second, it should be noted that what I provide here is merely a model; it supposes certain formal features of natural-language concepts which are indeed disputable, such as the supposition that employed scientific methods can universally be expressed as propositional axioms rules of inference. Such formal details are interesting to investigate and to put into question, but they are not the primary focus of the present discussion; I here intend to simply show that such a model provides an adequate representation of scientific change under somewhat simplified circumstances while still being properly contextualized; a proof of principle.

Another pertinent note is that, while possible worlds have been used already to study scientific change, for instance by Oddie and Niiniluoto (Oddie, 1986; Niiniluoto, 1987; Psillos, 1999), previous authors have explored the content of *particular* theories using possible worlds, and studied aspects of truth-likeness of certain theoretical claims *within* a particular theory via a sort of modal analysis. While I use some of these same tools in my analysis here, I deploy them at a completely different layer of analysis of the scientific enterprise. I here coarse grain away from the content of individual theories, taking *theories themselves* to essentially be the atomic symbols of the logical framework in question. Thus, the modality is one of theory acceptance by communities in light of their employed methods, and *not* one of the manner in which theories track truth in the world. Indeed, the framework developed here is closer to scientonomy, and perhaps even some sort of anti-realist leaning, formal social epistemology than it is to the sort of investigation of the logical structure of physical reality provided by the aforementioned authors in the past.

There have also been formal reconstructions of historically contextual theories of scientific change before as well, however, these do not extend far beyond reconstructions of Kuhnian concepts (Niiniluoto, 1981; Sadovsky, 1981). The formal framework provided here is not so much a reconstruction program as a parallel development to scientonomy, bearing in mind many of the lessons from integrated HPS since the 1960s.

Without further ado, I now present a formal account of scientific theory change using the above consideration.

First, while there has been extensive literature discussing the formal representation of theories, for instance, in terms of logical models or categories (Suppes, 1960; van Fraassen, 1980; Suppe, 1989; Glymour, 2013; Halvorson, 2012, 2016, 2017; Lutz, 2017), and through these discussions, many nuanced features of the equivalence of theories and their representations have become clear (see, for instance, Barrett, 2019), I here take a very minimal, course-grained view of theories. Specifically, I consider theories to be single, atomic propositions (these will carry with them a binary value of *acceptance*, rather than the usual *truth*). The idea here is that complex, full-blown theories (think quantum mechanics or genetics) may be viewed as the connective constructions (e.g., conjunctions) of smaller 'atomic' pieces. I therefore take the following definition.

> **Definition:** Let $\mathcal{T}$ denote the set of atomic theories of the world. Then the language of theories is given by $\mathcal{L} = \{\wedge, \vee, \rightarrow\} \cup \mathcal{T}$, where each $\tau \in \mathcal{T}$ is understood to be an atomic symbol.

This primitive language will be used in all models of theory change described here. The generic model I shall define will serve as a model for an arbitrary *particular* theory change, transitioning out of an arbitrary *pre-specified* mosaic. I call such a theory change an *episode*, and will provide a more explicit definition shortly.

To this language $\mathcal{L}$, we wish to include n-place predicate symbols. The episode of scientific change under consideration will dictate *which* predicates are pertinent to include. However, we may generically write $\{\Pi_n\}$ to be an arbitrary collection of n-place predicate symbols (where n is not fixed), and then specify them by context. Then the (unquantified) predicate language I shall consider below is $\mathcal{L}^* = \mathcal{L} \cup \{\Pi_n\}$.

To proceed, I shall let $\mathcal{M}_{C,t}$ denote the mosaic of community[1] C at time t. A mosaic specifies a collection of accepted theories and employed methods.

The class of all mosaics shall be denoted $\mathbb{M}$. Given a particular mosaic $\mathcal{M}_{C,t}$, we may define $\mathbb{M}_{C,t}$ to be the subset of $\mathbb{M}$ which have the same employed methods as $\mathcal{M}_{C,t}$, and which are defined for a time greater than or equal to t and a community which may, in some way, be identified with C. To each mosaic,

---

[1] There is an interesting question about the nature of indexing social communities, which are presumably fluid entities; such a problem of social indexicals, however, I do not explore here.

there shall be an associated world which I shall denote by the same symbol. Then any theory change takes $\mathcal{M}_{C,t}$ to an accessible mosaic in $\mathbb{M}_{C,t}$

Individual communities take stances of acceptance or unacceptance towards theories in $\mathcal{T}$. Thus, for a particular community C at time t, there is a function $\mathcal{I}_{C,t}: \mathcal{T} \rightarrow \{0,1\}$ which assigns an acceptance value to each theory $\tau \in \mathcal{T}$ (i.e. '0' represent a theory being unaccepted, and '1' represents a theory being accepted). The map $\mathcal{I}_{C,t}$ represents the theories which the community accepts at time t. I note explicitly that $\mathbb{M}_{C,t}$ is just the collection of mosaics for communities which are essentially C at later times which take possibly different stances of acceptance to different theories, and thus are may have a *different* interpretation $\mathcal{I}'_{C,t}$.

For a particular mosaic, the acceptance interpretation map thus described is not an employed method. It is more or less a historical artefact, which, for any given mosaic, we may presume to be given from the outset of our analysis. With this in mind, we may define an episode as follows:

Definition: An episode $E_{C,t}$ is a triple $E_{C,t} = \langle \mathcal{M}_{C,t}, \mathcal{I}_{C,t}, \{\Pi_n\} \rangle$, where $\{\Pi_n\}$ includes all of the predicates necessary to formulate the methods of $\mathcal{M}_{C,t}$ and $\mathcal{I}_{C,t}$ is the interpretation at the mosaic $\mathcal{M}_{C,t}$.

For the episode $E_{C,t}$, the mosaic $\mathcal{M}_{C,t}$ shall be referred to as the *base mosaic*. In what is to follow, for a given episode $E_{C,t}$, I construct a model of modal propositional logic (with unquantified predication) such that all mosaics which differ from the base mosaic by *logically* possible theory changes (i.e $\mathbb{M}_{C,t}$) are represented as possible worlds, but only those worlds (i.e., mosaics) which differ from the base mosaic in a manner which coheres with the methods of that base mosaic are *accessible* to the base mosaic. Then the class of *scientifically* possible theory changes correspond to the collection of mosaics which are accessible to the base mosaic.

In order to do this, it is necessary that I specify three things; (i) the *axioms* of the logic of a given episode, (ii) the *rules of inference* of the logic of a given episode, and (iii) the *accessibility relation* between mosaics in a given episode.

In principle, the axioms could be taken to be episode-dependent and variable. However, in this setting, axioms essentially just tell us what strings of symbols in the language of atomic theories (which are then understood to be more complex theories) are trivial. For instance, given a theory $\tau \in \mathcal{T}$, $\tau \rightarrow \tau$ is also a theory (and, in classical logic, $\vdash \phi \rightarrow \phi$ is generally taken to be an axiom). However, it is a trivial one, for if we evaluate it, we see that it is always accepted. In this way, the underlying axioms of the system are uninteresting; while they *may* be historically contingent, I suspect that the deviation from classical axioms will generally be minimal. Thus, for my purposes here, I simply suppose them to be the usual Hilbert-Ackermann axioms for classical logic.

I now turn to episodic rules of inference, which are far more interesting here. In logic, a rule of inference is a pair (Σ, ϕ) where Σ is a set of formulas. This pair is interpreted as saying $\Sigma \vdash \phi$; that is, from Σ, ϕ may be deduced. For the model being constructed here, the rules of inference will be the usual modal propositional logic rules of inference (i.e., modus ponens and the necessitation rule, see the Appendix for details), together with (possibly predicated) rules of inference due to the employed methods of the base mosaic. Before making this precise, I briefly describe why this is a natural way to formalize scientific methods in the present setting.

Scientific methods occur at all levels of scientific practice, ranging from theory formulation and justification, to experimental procedures, to the interpretation of experimental observations (Nola, 2007). However, all methods in all aspects of science serve as conditional protocols for action; experimental methods, for instance, may stipulate that *if* one is measuring a particular kind of system in a particular way, *then* they ought to do so in a particular (possibly implicitly) prescribed manner. Likewise, methods which are used for theory assessment (these are the only methods considered by scientonomy, and are also the only ones discussed here) stipulate that *if* a theory exhibits certain qualities or bears certain relations to other accepted theories, *then* that theory is also accepted or, and this is key, *could in principle become accepted*; I shall refer to methods of the latter type as *modal* methods. In this way, methods exhibit the sort of conditionality characteristic of rules of inference. I make this more precise.

In the framework this laid out, scientific methods for theory assessment are expressions which say 'from theories Σ, the theory ϕ may be deduced'. Using the acceptance-based semantics of modal propositional logic described here, this then entails that if all theories in Σ are accepted, so too is ϕ accepted as the theory ϕ is viewed as a logical consequence of Σ. However, if $\phi := \Diamond \psi$, then such a rule of inference indicates that, in light of the acceptance of Σ, ψ is *acceptable*. It is this modality which leads to the denotation of 'modal methods'.

Allow me, now, to elaborate just how predicates become important here. Frequently, methods are not just concerned with the abstract acceptance of theories, but rather, the properties these theories have or otherwise bear in relation to other theories. For instance, in hypothetical-deductivist episode, we may wish to say that one employed method is of the form 'if theory τ makes novel predictions, and if these novel predictions have experimental confirmation, then τ is acceptable'. The import of notions of 'novel prediction' and 'experimental verification' are not easily captured in a merely propositional logic. However, they may be captured by a clever choice of predicate. For instance, we may take CN to be a one-place predicate which are interpreted by reading CNτ as ' τ makes novel predictions which are confirmed by experiment'. Then the above hypothetical-deductivist method may be expressed as a rule of inference of the form (CNτ, $\Diamond$τ).

I claim that all methods in a given episode (i.e., the methods of the base mosaic) may be expressed in this way as additional rules of inference which are either modal or non-modal. The non-modal methods shall be added to the logic of a given episode as rules of inference. The modal methods, however, shall be used to determine an accessibility relation between mosaics in the possible-world structure of a given episode. Specifically, if we take $\mathcal{I}$ to be an interpretation function which assigns every possible interpretation to some element of $\mathbb{M}_{C,t}$, reducing to $\mathcal{I}_{C,t}$ on $\mathcal{M}_{C,t}$, I take the accessibility relation $\mathcal{R}$ to be defined by:

> **Definition:** A mosaic $\mathcal{M}'$ (for a community C' and time t') is accessible to the mosaic $\mathcal{M}_{C,t}$, denoted $\mathcal{M}_{C,t} \leq \mathcal{M}'$ (i.e. $\langle \mathcal{M}_{C,t}, \mathcal{M}' \rangle \in \mathcal{R}$) if and only if (i) $t \leq t'$ (ii) the community C' may be identified[2] with C, (iii) and if $V_{\mathcal{I}}(\phi, \mathcal{M}') = 1$ and $V_{\mathcal{I}}(\phi, \mathcal{M}_{C,t}) = 0$, then there is an employed modal method $(\Sigma, \Diamond \phi)$ in the mosaic $\mathcal{M}_{C,t}$ and additionally, $V_{\mathcal{I}}(\Sigma, \mathcal{M}_{C,t}) = 1$.

It is worth recalling once again that it was assumed that methods do not change across this transition; this is sufficient for theory change scenarios, but is obviously inadequate for method change scenarios. Condition (i) is to ensure that mosaics are historically ordered. Condition (ii) is to ensure that accessibility reflects changes in a communally shared belief system. Condition (iii), where the meat of the definition lies, indicates that two mosaics are only accessible insofar as, if the latter introduces a *new* theory (while keeping methods fixed), this new theory was acceptable as per the employed methods of the former.

We immediately see that $\mathcal{R}$ is reflexive, for whenever $C = C'$ and $t = t'$, only the third condition needs to be satisfied, but this is trivially satisfied, as the antecedent is always false. However, the requirement that $t \leq t'$ ensures that $\mathcal{R}$ is not symmetric. It should also be noted that, unlike in scientonomy where only single theory changes are allowed at a time, here, multiple theories may become accepted in a single shot.

---

[2] I am being intentionally vague here as community identification is notably challenging to make precise.

In relation to scientonomy, we have the following result built into this framework, assuming conclusive assessment outcomes. First, we note that the modal methods studied in scientonomy are called acceptance criteria.

Proposition (Second Law): A theory τ may become accepted by a community C at time t only if it satisfies the acceptance criteria of C at t.

Proof: If τ satisfies the acceptance criteria of C at time t, this means that there is an employed modal method $(\Sigma, \Diamond\tau)$ such that $V_{\mathcal{I}}(\Sigma, \mathcal{M}_{C,t}) = 1$. Likewise, if τ becomes accepted, this means that τ is not accepted in the mosaic $\mathcal{M}_{C,t}$, whence $V_{\mathcal{I}}(\tau, \mathcal{M}_{C,t}) = 0$, meanwhile, there is an accessible mosaic $\mathcal{M}'$ with $V_{\mathcal{I}}(\phi, \mathcal{M}') = 1$ at some $t' \geq t$. By the definition of the accessibility relation, we therefore see that acceptance of τ may only occur if it satisfies the acceptance criteria of C at t.

This result resembles the second law from scientonomy. We therefore see that this model recovers much of the substantive content of theory change present in scientonomy.

The full model of theory change, as developed here, is as follows. First, one specifies a particular episode $E_{C,t}$ with an associated community C, at time t. This episode has associated with it a base mosaic $\mathcal{M}_{C,t}$ which comes equipped with a collection of methods, both modal and non-modal. There is also a collection of predicates which must be specified in order to formalize these methods.

Associated with the episode $E_{C,t}$, there is an extended (predicated but unquantified) propositional logic which includes the formalized non-modal methods as additional rules of inference, and a logical model of modal logic which takes this episodic logic as its base. The associated model is a triple $\langle \mathbb{M}_{C,t}, \mathcal{R}, \mathcal{I} \rangle$ where $\mathbb{M}_{C,t}$ denotes a collection of possible worlds (interpreted as mosaics which share the same methods as $\mathcal{M}_{C,t}$), $\mathcal{I}$ is the extended acceptance interpretation function which reduces to $\mathcal{I}_{C,t}$ on the base mosaic world $\mathcal{M}_{C,t}$ and $\mathcal{R}$ is an inter-world accessibility relation defined above using the modal methods of $\mathcal{M}_{C,t}$

Using this framework, the collection of permissible scientific theory changes which the mosaic $\mathcal{M}_{C,t}$ may undergo is given by looking at the acceptance interpretation function on each mosaic which is accessible to $\mathcal{M}_{C,t}$. That is, each accessible mosaic to $\mathcal{M}_{C,t}$, with accepted theories given by the restriction of $\mathcal{I}$ to that particular world, is a possible future state into which the community C may scientifically transition.

3. Episodic Rationality

There is a natural worry that, by overly formalizing a theory of scientific change in terms of some precise deductive logic, one is erasing the *human* elements of science which have been the subject STS and HPS for several decades now. I hope to now put the socially or historically oriented reader's mind at ease.

While the framework I have developed here does provide an underlying logic for studying scientific change – one which is indeed deductive – thereby providing a notion of rationality of scientific change in each individual episode of change, the model itself is *vastly* underdetermined. To see this, we need only realize that the model only indicates which scientific changes are *permissible*; the modality of the framework is intentionally not deterministic. The account above allows a mosaic to transition into any *accessible* mosaic, however, it provides no indication of *which* mosaic it will actually transition to. Thus, it acts more as a logical *constraint* on admissible scientific changes.

In light of this underdetermination, there is now an obvious place for sociocultural factors to play a substantive role in scientific change. Socially contingent human agency is responsible, not for deciding what changes are admissible by their employed framework in some instant; this would allow *arbitrary* theory changes to take place. Rather, this contingent agency of the scientific community plays the role in *deciding* which accessible mosaic should become *actual*. In this manner, science is simultaneously a rational and contingent human enterprise because, on the one hand, every scientific change obtains within the logical constraints of a particular belief system (hence, there is a local notion of rationality). However, the choice of which allowable direction to go is *not* one which is logically prescribed, and is crucially dependent upon the actual human practice of science.

Additionally, while I have only here dealt with *theory* change, *method* change is also possible (in Section 4, I discuss the possibility of extending this framework to include method change as well). However, the episodic logic upon which theory change is based is rooted in the methods employed in the base mosaic of that episode. Thus, when method change occurs, so too does the underlying logic of scientific change. In this way, the logical *direction* of change is constantly reorienting itself, making it impossible to define a global logic (and thus a global notion of scientific rationality) which scientific change respects. It is, however, possible, as I have here demonstrated, to understand any *particular* theory change (and, in principle, method change as well) in terms of some episodic logic. Thus, there is a *local* notion of rationality which may be recovered. I call this phenomenon *episodic rationality*.

To see how this notion of episodic rationality respects social contingency, consider the following: in any scientific mosaic at a particular time, there will likely be a large class of *possible* theories which, if constructed properly would satisfy the acceptance criteria of the community at that time and stand a reasonable chance of becoming accepted. However, scientific theorizing is, at the most basic level, contingent upon which theories particular scientists choose to explore. That is, there is an accidental constraint on *actual* scientific change due to the contingent factors in the context of discovery. Likewise, even if an acceptable theory is proposed, if it is inadequately presented -- by being too confusing and poorly derived for example --, it may still not become accepted.[3] Even if a theory τ_1 is thought of and formulated in an agreeable manner such that it is amenable to acceptance, there may happen to be a competing theory τ_2 on offer which is in some way preferable to τ_1 for the scientific community (for instance, by its simplicity, or readiness to recover experimental observations, or its superior prediction-making ability, etc.). In such an instance, τ_2 may become accepted while τ_1 remains unaccepted *even though* τ_1 is otherwise acceptable and *would have* become accepted if τ_2 had never been formulated.

This line of reasoning quickly becomes obscure, as any concrete example of this relies on a counterfactual history of events which never occurred. However, the message is clear; science is contingent, and this contingency is closely coupled to sociocultural factors. However, it is these factors which steer scientific change in one direction over another in light of the underdetermination implicit in the underlying logic. In this way, the ramifications of contingent social effects on science, while unpredictable and not necessarily logically motivated, are still confined to a domain of episodically rational change. Thus, each instant of scientific change is rational *qua* the logic underlying the worldview within which the assessment takes place.

4. Discussion

One criticism of the scientonomic project is its limited scope of scientific methods and values. Indeed, many features of science which exhibit particular methods are not studied by scientonomy; as Chang noted (Chang, 2022), scientonomy is perhaps too course-grained to accommodate interesting features of science, such as experimentation, at least in its current state. One of the benefits of the model provided above is that it may readily be extended to novel

---

[3] Consider, for example, the recent instance of Mochizuki's purported proof of the ABC conjecture using his so-called 'trans-universal Teichmüller theory'. This theory is so convoluted that the mathematics community remains unsure whether or not the demonstration of the theory satisfies their acceptance criteria, which may be thought of as proof conditions in classical logic (Castelvecchi, 2020).

contexts by the introduction of additional layers of logic structure. For example, in order to recover full scientonomy, one could, in principle, carry out an identical analysis as provided here at the *meta-logical* level, taking methods to be certain forms of modal conditional formula schemas (with the syntactic conditional of the higher-level logic replacing the semantic entailment of the one thus described), obtaining a robust formal account of *method* changes as well.

This framework may also be extended in a wide range of other ways as well. One could, for instance, specify particular n-place predicates of interest (e.g., simplicity or falsifiability as 1-place predicates, compatibility as a 2-place predicate, and whether or not one theory is reducible to a collection of n other theories as an (n + 1)-place predicate, and so on) and investigate their relation with respect to the acceptance-based semantics described here.

Another option is to introduce new modal operators representing different stances scientific communities may have towards theories (just as knowledge or belief may be modelled as a modal operator in epistemic logic and doxastic logic, respectively (Hintikka, 1962; Sider, 2010)). In doing so, additional sorts of scientific stances towards theories could be reflected.

In principle, this framework could be extended to a multidimensional modal semantics (Marx, 1997) to simultaneously encode a variety of different properties of scientific theories and our stances towards them. Likewise, as the meta-logical level, all of this may be done for methods as well. Multiple accessibility relations could also be introduced to overlay multiple different modal structures simultaneously, for instance, using concurrency theory (e.g., labelled transition systems). It is also conceivable that framework could be recreated with different terms (i.e., not theories, but, say, experimental observations) with semantics grounded in different values beyond 'acceptance' to encapsulate more features of the scientific enterprise (ideally finding a manner to stitch it all together coherently).

Prior to such extensions, even within the scientonomy setting, it remains an open question[4] whether or not all communities are capable of assessing the acceptability of all historical theories. Indeed, the case could be made that incommensurability may be so severe across large historical and cultural distances that certain communities simply cannot assess the acceptability of

---

[4] Though not presently formulated in the encyclopedia, this issue is related to the 'Status of Tacit Theories' open question (https://www.scientowiki.com/Status_of_Ta cit_Theories); rather than asking if a community can accept a theory that isn't formulated, it is asking if they are able to counterfactual assess theories which are formulated, but not present in their own mosaic (see Fraser & Sarwar, 2018, p. 70, footnote, for details).

certain theories. In the present article, this translates to the possibility that the interpretation functions $\mathcal{J}_{C,t}$ might not be defined for all theories $\tau \in \mathcal{T}$. To avoid such complications, the present framework may be modified using variable domains to restrict the space of theory formulas a mosaic will encode information about. The modal framework is sufficiently rich that it provides many avenues for resolving or accommodating such difficulties which may, in plain-language philosophy, be obscure or difficult to fully appreciate and control.

5. Conclusions

Here, I have provided an analytical model for the logical analysis of scientific theory change. This model was motivated by the notion of a mosaic from scientonomy and the realization that mosaics may be interpreted as possible worlds under some accessibility relation. I made this notion precise, and introduced the notion of an episodic logic of scientific theory change which takes as rules of inference the (predicated) formalization of employed methods in some base mosaic. I then showed that a robust account of theory change, which recovers the second law from scientonomy, may be developed using these tools. Such a model exhibits what I have here called *episodic rationality*; each theory change is constrained to occur within some extension of classical logic, whence it is rational in a local sense. However, the underlying logic is, itself prone to change (via method change), whence there is no global notion of rationality available. The underdetermination of this model allows for an explicit, determining role for sociocultural and other contingent factors to play a substantive, though logically confined, role in scientific change. I then discussed the possibility of extending this model to encapsulate additional features of science beyond theory acceptance, such as experimentation, and hinted at the possible formal extensions which could make such an extended theory feasible.

Bibliography

Barrett, T. W. (2019). Equivalent and Inequivalent Formulations of Classical Mechanics. *British Journal for Philosophy of Science*, 70, 1167-1199.

Barseghyan, H. (2015). *The Laws of Scientific Change*. Springer.

Burgess, J. (2012). *Philosophical Logic*. Princeton University Press.

Carnap, R. (1937). *The Logical Syntax of Language*. Kegan Paul.

Castelvecchi, D. (2020). Mathematical Proof that Rocked Number Theory will be Published. *Nature*, 580(177).

Chang, H. (2022). The Ontology of Scientific Practice. In this volume, 1-19.

Feyerabend, P. (1975). *Against Method*. Verso.

Fraser, P. & Sarwar, A. (2018). A Compatibility Law and the Classification of Theory Change. *Scientonomy*, 2, 67-82.

Glymour, C. (2013). Theoretical Equivalence and the Semantic View of Theories. *Philosophy of Science*, 80 (2), 286-297.

Halvorson, H. (2012). What Scientific Theories Could Not Be. *Philosophy of Science*, 79 (2), 183-206.

Halvorson, H. (2016). Scientific Theories. In Humphreys (Ed.) (2016).

Halvorson, H. & Tsementzis, D. (2017). Categories of Scientific Theories. In Landry (Ed.) (2017), 402-429.

Hintikka, J. (1962). *Knowledge and Belief.* Cornell University Press.

Hintikka, J., Gruender, D., & Agazzi, E. (Eds.) (1981). *Theory Change, Ancient Axiomatics, and Galileo's Methodology: Proceedings of the 1978 Pisa Conference on the History and Philosophy of Science.* D. Reidel.

Humphreys, P. (Ed.) (2016). *The Oxford Handbook of Philosophy of Science.* Oxford University Press.

Kuhn, T. S. (1962). *The Structure of Scientific Revolutions.* University of Chicago Press.

Kuhn, T. S. (1982). Commensurability, Comparability, Communicability. *PSA: Proceedings of the Biennial Meeting of the Philosophy of Science Association*, 1982, 669-688.

Lakatos, I. (1970). Falsification and the Methodology of Scientific Research Programmes. In Lakatos (1978), 8-101.

Lakatos, I. (1978). *Philosophical Papers, Volume I.* Cambridge University Press.

Landry, E. (Ed.) (2017). *Categories for the Working Philosopher.* Oxford Scholarship.

Laudan, L. (1984). *Science and Values.* University of California Press.

Lutz, S. (2017). What Was the Syntax-Semantics Debate in the Philosophy of Science About? *Philosophy and Phenomenological Research*, 95, 319-352.

Marx, M. & Venema, Y. (1997). *Multi-Dimensional Modal Logic.* Springer.

Merton, R. K. (1973). *The Sociology of Science: Theoretical and Empirical Investigations.* University of Chicago Press.

Niiniluoto, I. (1981). The Growth of Theories: Comments on the Structuralist Approach. In Hintikka, Gruender, & Agazzi (Eds.) (1981), 3-47.

Niiniluoto, I. (1987). *Truthlikeness.* D. Reidel.

Nola, R., & Sankey, H. (2007). *Theories of Scientific Method.* Routledge.

Oddie, G. (1986). *Likeness to Truth.* Springer.

Psillos, S. (1999). *Scientific Realism: How Science Tracks Truth.* Psychology Press.

Quine, W. V. O. (1951). Two Dogmas of Empiricism. *The Philosophical Review*, 60, 20-43.

Sadovsky, V. N. (1981). Logic and the Theory of Scientific Change. In Hintikka, Gruender, & Agazzi (Eds.) (1981), 49-61.

Sider, T. (2010). *Logic for Philosophy.* Oxford University Press.

Suppe, F. (1989). *The Semantic Conception of Theories and Scientific Realism.* University of Illinois Press.

Suppes, P. (1960). A Comparison of the Meaning and Uses of Models in Mathematics and the Empirical Sciences. *Synthese*, 12, 287-301.

van Fraassen, B. (1980). *The Scientific Image.* Oxford University Press.

Appendix: Modal Logic

Here, I define the basic syntax and semantics of modal propositional logic (MPL) with its extension to include (unquantified) n-place predicates. In the model constructed in Section 2, I explore *extensions* to this logic by the introduction of additional rules of inference. For a more detailed treatment of MPL, the reader is referred to (Sider, 2010, ch. 6), which I follow closely here.

First, the syntax begins with a language $\mathcal{L}$ which consists of atomic symbols {p}, negation ¬, the connective → (all other standard connectives may be constructed using ¬ and →), a collection of n-place predicates {Π_n} for each n > 0, and modal operators ◇ (possibility), and □ := ¬◇¬ (necessity). A formula ϕ in this language is well-formed (it is a *wff*) if it is of the following form:

- ϕ is an atomic symbol

- ϕ is defined by ¬ψ or $\psi \rightarrow \chi$ for wffs ψ and χ

- ϕ is defined by $\Pi p_1 \dots p_n$ for some n -place predicate Π with each p_i an atomic symbol

- ϕ is defined by ◇ψ or □ψ for some wff ψ

The axioms of this system are those of ordinary propositional logic (e.g., the Hilbert-Ackermann axioms) together with the K axiom, ⊢ □($\phi \rightarrow \psi$) → (□$\phi \rightarrow$ □ψ). The rules of inference of this system are modus ponens which stipulates {$\phi, \phi \rightarrow \psi$} ⊢ ψ for all wffs ϕ and ψ, and the necessitation rule ϕ ⊢ □ϕ. A *deduction* $\Sigma \vdash \phi$ in this system, is given by a finite sequence of formulas where each string is either a member of Σ, an axiom, or is obtained from a previous formula in the sequence by the application of a rule of inference. (This will also be the case additional rules of inference are introduced in Section 2.)

The semantics of this logic is given in the following manner. A model M of this logic is a triple $\langle \mathcal{W}, \mathcal{R}, \mathcal{I} \rangle$. Here, $\mathcal{W}$ denotes a set of possible worlds (formal objects which correspond to distinct frames of logical analysis) and $\mathcal{R}$ is an accessibility relation between possible worlds in $\mathcal{W}$. Then $\mathcal{I}$ is a two-place function which takes in either an atomic symbol or an atomic predicate wff (i.e., a wff of the form $\Pi_n p_1 \dots p_n$ for some n-place predicate Π_n) and a world w ∈ $\mathcal{W}$ and outputs a truth value in {0,1}. Then $\mathcal{I}(p, w)$ is the truth value of the atomic symbol p at world w, and $\mathcal{I}(\Pi p_1 \dots p_n, w)$ is the truth value of the predicate Π with respect to the particular choice of atomic symbols p_i at world w.

Given such a model, the truth interpretation function $\mathcal{I}$ is enough to determine the full semantic truth valuation of arbitrary wffs using the valuation function $V_{\mathcal{I}}$ in the following way:

- $V_J(\neg\phi, w) = 1$ iff $V_J(\phi, w) = 0$

- $V_J(\phi \rightarrow \psi, w) = 1$ iff either $V_J(\phi, w) = 0$ or $V_J(\psi, w) = 1$

- $V_J(\Diamond\phi, w) = 1$ iff $V_J(\phi, u) = 1$ for some world $u \in W$ with $w \leq u$.

- $V_J(\Box\phi, w) = 1$ iff $V_J(\phi, u) = 1$ for every world $u \in W$ with $w \leq u$.

If $V_J(\phi, w) = 0$ at all worlds w in the model M, we may write this as $\vDash_M \phi$. Further notions of semantic entailment may be defined in the obvious manner (again, I refer one to Sider, 2010, ch. 6, for details).

I note that, while predicates have been included in this system, they essentially just play the role of a special sort of atomic proposition, since there is not quantification present. In this way, this logic is essentially the same as ordinary MPL.

Chapter 7

Thinking Big: The Science of Change and the Historicity of Scientific Method

Guillaume Dechauffour

Sorbonne Université

Abstract: Scientonomy seems to hold conflicting views about the historicity of the scientific method. On the one hand, it is said that scientific methods are immanent to scientific mosaics and therefore change through time. On the other hand, the distinction between substantive and procedural methods seems to suggest that there are transcendent, unchangeable methods. I argue that this contradiction can be resolved by re-evaluating the role of problems: by integrating problems as constitutive elements of scientific mosaics, scientonomy can work towards a theory of scientific change without relying on the presupposition that some normative aspects of science must not change. In that perspective, norms originate in the relation between a problem, which creates a need for theoretical innovation, and a method, which creates an actual means to solve a problem. A problem-based scientonomy would then have to build a genealogical, rather than normative, approach to the source of scientificity by describing the progression from mysteries to scientific problems. Moreover, because they do not come from nowhere but express actual interactions with the world, problems can help us understand the relation between scientific change and other kinds of change. The primacy of actual problems over rational norms points to the immanence of reason: reason should be conceived as an evolutive feature of human communities. Finally, the relation between a theory of scientific change, evolutionary epistemology, and a general theory of change is investigated.

Keywords: scientonomy; evolutionary epistemology; cosmic evolution; methodology; problems; scientific change

***

1. Introduction

The way in which open questions are clearly identified and publicized[1] is one of the great things about scientonomy. It enables newcomers to understand immediately how they could contribute to the field. Thus, as an introduction, I can easily pinpoint, three open questions I wish to address:

- Can the Theory of Scientific Change (TSC) be used to solve the problem of scientific progress?

- Are the laws of scientific change reducible to some sociological or psychological laws?

- What is the place of problems (issues, questions) in the scientific mosaic? What role do problems play in scientific change?

My general claim is that scientonomy, as a TSC, cannot ignore the existence of a scientific theory of change, or else it takes the risk to repeat an excessively idealistic conception of science as purely intellectual and autonomous process.

My contribution comes from an evolutionary perspective. Evolutionary epistemology pursues the idea of rebuilding philosophy of knowledge around the fact that knowledge is an evolutionary process, both at the individual and collective scales. Describing scientific change is therefore an essential part of this project. Thus, there is a natural connection between evolutionary epistemology and scientonomy. The main goal of evolutionary epistemology is to dissolve the tension between scientific inquiry and philosophy by admitting that there is nothing more to find than what science can describe, that is the everchanging process that shapes the world and the ways we may interact with it. As a result, there would be no reason to conceive of epistemology as a search for an absolute criterion of adequation between our theories and what they describe and no contradiction left between the variability and relativity of actual scientific research and this philosophical ideal. Starting from the actual history of science rather than a philosophical ideal, we must acknowledge as a basic fact that there will always be something new to know, because both the world and our scientific theories are changing realities, which is the very root of the problem of scientific progress. In that line of thinking, our epistemic lives are a part of the fractal structure of our evolving world, and epistemology must be articulated within a general science of change. So if we want to investigate the implication of a TSC regarding scientific progress, then we should consider

---

[1] https://www.scientowiki.com/List_of_Open_Questions

its relation not only to the evolution of our collective capacity to comprehend reality, i.e. evolutionary epistemology, but also to a general theory of change.

Currently, the most notable scientific framework which takes universal change as its object is *Cosmic Evolution* (Chaisson, 2001; Shaver, 2011), hereafter *CE*, which goes hand-in-hand with the *Big History* pedagogical program, hereafter *BH* (Cloud, 1978; Christian, 2004). Surprisingly, neither CE nor BH include scientific change as a part of the change they study. They generally stop their inquiry or narrative at the emergence of intelligent behavior or in some cases at the evolution of human societies. Perhaps by fear of circularity, they do not go as far as to include themselves in their subject matter by addressing the evolution of techno-scientific cultures. Conversely, Scientonomy does not say much about the relation between scientific change and other kind of transformative processes. But to understand change, we must *think big* (Gamble, Gowlett, & Dunbar, 2018): we must take as much distance as we can for patterns to appear, so that we would be able to connect the different scales and processes involved in the evolution of our understanding of the world. Thus, we must be able to embrace every kind of change, both material, practical and intellectual, in a synoptic view.

Reciprocally, if bigger is better, then we must understand what it means to scientifically go bigger: how, as science comes to acknowledge the chronological features of its objects (Toulmin & Goodfield, 1982), a historical, genealogical, or mobilistic methodological turn must happen that will allow a superior transdisciplinary theoretical coherence. This is precisely the kind of transformation scientonomy aims to describe. I hope it has become obvious, through this brief analysis of the problem of scientific progress, that scientonomy and CE have a lot to offer to each other and must be combined in order to determine what *change* is and how it should be studied.

Another reason why most champions of CE ignore scientific change might be that they subscribe to a rather naïve view of science, according to which science is only the application of universally available rational tools to study objective features of reality. It is a convenient view, because it gives evolution an unquestionably objective status since it is the result of a scientific inquiry and reciprocally, it gives science an unquestionable status as the end result of general evolution. But it is also a rather paradoxical view to hold if one believes that everything is always changing. Why would science, which has manifested constant evolution in its every aspect throughout history, not be part of the general movement? Therefore, highlighting the historicity of scientific methods is the right starting point to demonstrate the major contributions that scientonomy can make to a general theory of change. This is where we will meet the second question.

But this contribution is also difficult from a scientonomic point of view, as there may be a tension between the way scientonomy conceives the changeability of method and the way it describes the necessary conditions for scientific change. I think this could be solved by including problems, understood as the source of the drive for change, in these necessary conditions and, since problems do not only refer to theoretical puzzles expressed by questions but also concrete difficulties in practical situations, they can also assume the function of articulating the theory of scientific change and a general theory of concrete and practical change. This is where we will meet the third question.

In scientonomic terms, I will try to present a version of the argument *from nothing permanent* (Barseghyan, 2015, p. 83), applied to method, but my conclusion is not the impossibility of a TSC but the necessity of articulating it to general theory of change through a genealogy of problem, so that it is more akin to an argument *from nothing transcendent.* My main points are that if science is a problem-solving activity and that if there are some permanent features of science, they must come from the fact that some problems remain unsolved and, since problems can be genealogically tracked back to concrete difficulties, the permanent features of science can also be tracked back to concrete difficulties experienced in practical contexts, that is to environmental constraints on some scale of the general evolutionary process.

2. The Changeability of Method

The definition of method is the first task Hakob Barseghyan considers in *The Laws of Scientific Change* (Barseghyan, 2015). I have identified four steps or sub-thesis in this definition and I will show how and when a philosophical tension seems to appear regarding the historicity of science. Step 1 stipulates that *methods are theoretical norms and criteria of theory assessment.*

Barseghyan (2015, pp. 4-5) clears the traditional ambiguity surrounding the concept of method. The word "method" can refer to the way things are done or the way they should be. In a scientific context, it refers whether to the way theories are invented or to the way they are evaluated and justified. The confusion of those meanings is unfortunate, since it tends to conceal the selective nature of scientific change, the fact that scientific communities aim to select the best theory available. Therefore, Barseghyan choses to use the word "method" only in the normative sense, as "a set of criteria for employment in theory assessment" (Barseghyan, 2015, pp. 4-5).

This is the first step; its goal is to establish that methods are not simply practical guidelines on how to do things, they rather edict norms about what a theory should be to be accepted. The reason behind this semantic choice is that science has a normative dimension which must be accounted for. This normative dimension is twofold: it concerns both the exercise of science and its results. First, it implies that one cannot do science as one pleases, there are certain ways of observing, reasoning, and formulating theories which are valid and others which are not. But it also has to do with the status of these theories. If they have been produced according to the valid procedures of science, then they become endowed with normativity: theories assert a higher claim to truth and any idea that contradict established theories is doubtful. That is in part where the problem lies: we need norms in order to accept a theory and we need our theories to have some authority. And if these norms are relative or unstable, the condition for the collective use of reason and the collective acquisition of theories do not seem to be met. Therefore, the very existence of science and scientific theories in a very diverse world seems to entail that norms must apply necessarily. The distribution of normative power between methods and descriptive theories has been addressed in the most recent update of the scientonomic framework (Barseghyan, 2018), where methods are conceived as normative theories.

Step 2 suggests that there is no difference between method and value. Further in the same inaugural subchapter, Barseghyan argues that values are expressions of our goals and methods formulate requirements, and since goals can be seen as requirements and vice versa, any value can be adequately understood as a method and every method convey values.

The permutability of method and values in different accounts of science allows us to conclude that there is no difference between them. In other terms, there is no difference between categorical and hypothetical imperative in science. If everything is a method and nothing stands above, then one could conclude that there are only hypothetical imperatives, ways to achieve a goal, and no architectonic values which would justify on a higher level the methods we chose to apply.

Step 3 stipulates that there is no unchangeable scientific method. In the next subchapter, the idea of an unchanging method is falsified by historical evidence: contemporary scientists do not conduct research in any way like the Renaissance polymaths did. There are two consequences to this undeniable fact: firstly, methods are part of the evolving scientific mosaics scientologists aim to describe; secondly, it is absurd to pursue the search for an absolute method.

It is this claim that got me excited about scientonomy. Firstly, because this idea is remarkably close to the axiom of evolutionary epistemology: methods are not universal and transcendent but diverse and historical. But if we also take step 2 into account, it follows that this changeability applies to all scientific norms and values, including our conceptions of truth or validity. The consequence is gigantic: a pure epistemology, which would aim to discover an unchangeable way to access some eternal and necessary knowledge, is pointless. Hence it means that epistemology must reconceive its own task and nature: it must aim to describe how new epistemic norms and goals emerge and come to replace older ones so that science itself, like the ship of Theseus, persist by constantly changing. This is precisely where scientonomy reveals itself as the meeting point between epistemology and evolutionary thought in general.

Though, from a scientonomic point of view, there is a more direct consequence: step 4, which stipulates that TSC and methodology (MTD) do not coincide. Indeed, if there are no eternal and transcendent scientific methods and values, then it seems that the study of the evolution of scientific practices and theories is different from the task of normative methodology. This is not only a conceptual point: the immediate consequence of distinguishing between a descriptive inquiry about change in theories and methods and a normative methodology is that we must henceforth conduct separately two tasks we used to conflate.

Barseghyan insists on this point, it is where he differs from former theories of scientific change and traditional philosophy of science in general. He prescribes a change of goal: since there is no unchangeable method, TSC should be conceived only as a descriptive theory. We must not aim to provide guidance to other researchers but turn our focus on explaining how scientific theories happen to change. As a result, the autonomy of scientific methodology must be respected. According to the scientonomic point of view, it is not the same thing to understand how scientific progress happens and to know how to make it happen.

It is important to note that this is also a redefinition of the scope and reach of methodology which completes what was engaged in step 3. By separating from a descriptive inquiry such as TSC, which ceases to be seen as normative, MTD can become really prescriptive. Yet, at the same time, its prescriptions become relative: methodology is not to be understood as the discovery of unchanging norms but as an effort to determine particular goals and ways to achieve them. If this is true, then the distinction between the conception of methods seen in 1, as guidelines and as norms, becomes blurrier, at least for MTD. We could say that TSC describes the evolution of methods as they apply to theories inside particular scientific mosaics but MTD develops and elaborates new and different ways of doing science and not a constraining definition of science. As a result, MTD seems to provide guidance rather than norms, even in the field of theory assessment.

As I wholeheartedly share the belief that TSC and MTD are different endeavors, I think nevertheless that a problem may arise at this point and that the resolution of this problem could be the contribution of a scientific theory of change to a theory of scientific change.

3. Baseless Methodology?

The scientonomic redefinition of *method* and epistemology leaves us with some ground-breaking ideas:

- Scientific methods are normative but diverse, they can change.

- There are no absolute scientific values which supervene over methods, they too can change.

- TSC cannot function as MTD: describing the evolution of norms is not a normative endeavor.

It also leaves us with a problem: if there are no definitive scientific values, goals, or categorical imperatives, on what basis should methodology operate? If it is on the basis of science history, without any evaluative element, then science is not different from tradition: we must consider valid what has been considered valid for the longest time. That is precisely what Bacon and all his followers argued against when they invented the modern scientific mindset and it is also something we immediately know to be wrong, since if it were not there would never be any meaningful change in science, and we would not be having this conversation in the first place.

At the end of the second subchapter, Barseghyan argues that, even if they are to be distinguished, MTD could find its missing basis in TSC (Barseghyan, 2015, p. 21):

> Let us consider both options. If it turns out that MTD cannot legitimately use our knowledge about the workings of science, this will mean that TSC has no bearing on MTD. But even if it turns out that descriptive TSC could be legitimately used in settling the issues of normative MTD, it will change nothing in our understanding of the scope of TSC itself. For to say that MTD should base its prescriptions on the findings of TSC is analogous to saying that the engineer should draw heavily on existing physical theories when deciding what material to use for building a bridge. This does not, however, make physics and engineering identical. In particular, the choice of the proper material for bridge building does not become a task of physical theory. The same relation holds between TSC and MTD. In deciding what method is to be employed in theory assessment, normative MTD may draw upon our theory of how science operates (TSC, that is), but it doesn't make TSC and MTD identical.

If we consider both options then, as already mentioned, we seem to face a problem. In the first option, that is if MTD cannot use TSC as its base, obviously we fail to see on what other base methodology could operate. It may be the case that the scientific method is not something that can be determined autonomously but rather in the context of actual scientific practice.

In the second option, MTD can operate based on TSC success. It means that one could edict norms of theory assessment using a description of how theories and methods come to be accepted by a scientific community. If that is the case, we might face two problems instead of one. First, if TSC could produce determining principles for methodology, the distinction would become quite blurry: it seems to me that TSC would function as methodology, the only difference being that the method of methodology would be empirical rather than *a priori*. The latter is a stimulating idea and a proper conclusion, but not the one actually drawn in the text. It might very well not be the case though, for it is difficult to see what kind of methodological principle could be drawn from TSC, as TSC does not provide any evaluation of methods: it doesn't discriminate between efficient or inefficient methods. To clarify this relation, Barseghyan uses an analogy which compares the relation between bridge-building and physics to the relation between MTD and TSC. But I do not think that these relations are similar.

The bridge-building/physics relation describes a connection between an empirical exploration of the material world and a technical practice which aim to produce efficient and sturdy structures. MTD on the other hand is not a technical practice, it does not actually build a particular thing that someone could use. It aims to edict rules. If the idea is that TSC, as an empirical description of general laws governing a domain of the external world, can provide guidance to the practical endeavours which involve this domain, then the second term of the analogy should not be methodology, which is an intermediary step, but an actual theory building: physics for example. It is difficult to see, moreover, how knowing the general laws of theory and method acceptance could determine the content of any scientific theory the same way physics can determine the materials used in bridge building.

The analogy between the first terms of each of its relations is also problematic. In what sense could TSC be similar to physics? Both are empirical inquiries and they both produce descriptive theories. But here again, the differences seem to outweigh the similarities. First, a successful theory in physics is a theory which makes accurate predictions possible, that is why it may be useful for an engineer to know physics in order to predict the behavior of a building depending on any given variable, whereas a successful theory of scientific change cannot predict when or how a part of a scientific mosaic will change.[2] Secondly, unlike physics, TSC's subject matter is not the ultimate structure of the outside world but to produce a model of our own activity. It is a reflective rather than an exploratory inquiry. Comparing TSC to physics tends to conceal the reflectivity of TSC, which is a science about science, and even its circularity, if TSC is both a description and a condition of scientific change. These two features are essential to TSC and are also the reason why it is difficult to consider TSC as a basis for MTD. Finally, if the relation between physics and bridge-building is interpreted as a model for the type of basis which scientific practice needs, then TSC is not what we should be looking at. The physics/bridge building exemplifies the relation between the knowledge of the nature of matter and energy and their general laws on the one hand and our ability to produce adequate material artefacts on the other. Then if we apply the symmetry, what scientific practice, understood as the production of adequate intellectual artefacts, really needs is knowledge of the nature of reality and ideas and their general laws, which is not what TSC aims to describe, it is rather the domain of metaphysics. And I do not think that the goal of scientonomy is to restore the role of metaphysics as *prima philosophia* or even to inherit this role. Another possible candidate would be a general theory of scientific rationality. But we have already rejected the idea of transcendent, universal, and unchangeable knowledge of norms and values. As a result, when we try to identify what could assume, regarding to scientific methodology, the role that physics plays for engineering, we seem to come up empty-handed.

Indeed, TSC is not to MTD what physics would be to bridge-building or engineering. It seems it would be what a general theory of technological progress would be to engineering: it can define what is an invention, what is a method of production, what is a technical domain and how it evolves. And even though a general theory of technological progress can describe the general laws of technical innovation and how it comes to change an entire domain of production or the activity of a given society, it would not teach anybody how to invent anything.

Hence, whether MTD finds its principles in TSC or not, the result seems to be the same: if we accept that methods and values are changeable, then

---

[2] It can and should, of course, try to predict what will happen to a given mosaic given such-and-such accepted theories and employed methods and given such-and-such competitors and body of evidence. What it cannot do is predict what the next stage of evolution will be or what will trigger it, since theory construction itself is a creative process. In other words, TSC cannot discover the next scientific discoveries before they are actually discovered, just like evolutionary biology cannot predict the apparition of new species and their full genetic profile (and even more so, since scientific change is determined by more diverse variables).

methodology cannot operate on any other basis than the scientific practice itself. It appears clearly, then, that the set of assertions described earlier as the redefinition of method do not entail that a TSC should be justified by its methodological utility, but rather than its philosophical consequences are closer to the pragmatist coherence advocated by Hasok Chang (Chang, 2016, 2022), if we think that systems of practice are inherently coherent, or even some sort of epistemic anarchism if we do not.

These philosophical positions are both sustainable and interesting, but this result leaves us facing the problem of normativity: if scientific methods and values are changeable, how can we decide if a theory is invalid or if we need to update our methodological values? Without a theory of what science should be, methodology seems baseless. One way of showing that is to consider the fact that demarcation criteria are a part of a method. How can we even decide if a theory or a method is scientific? If a method is founded on a prior scientific practice, as per the third law, then what does certify the scientificity of this practice if not another method? And what constitutes the scientificity of that method in the first place?

Therefore, if we want to reject the idea of an eternal and transcendent norm, we have to undertake a genealogy of scientificity which would not rely on any kind of self-validating procedure, or else we would make both a conceptual mistake, by understanding science as a self-produced and self-sufficient process and a logical one by creating the risk of an infinite regress.

4. Ambiguity

It appears that Scientonomy has yet to clearly make this choice, as the distinction between dynamic or static and substantive or procedural methods, in Part II, chapter 5, looks a little bit like backpedaling on this matter. Barseghyan introduces a distinction between two types of scientific methods. The first ones are the result of actual scientific practice and are logically linked to particular scientific propositions, they are called substantive and are subject to change. The procedural methods, on the other hand, are said to be independent of any scientific discovery or presupposition apart from necessary truths which are themselves independent of any historical process. In other words, procedural methods are scientific principles.

If there are methods which seem independent from the world we live in, we could infer that they are transcendent, but it could also mean that they are arbitrary. Then, we must figure out where these methods come from. To that end, we can draw a distinction between methods which presuppose contingent theories and methods which rely on necessary truths. Methods which rely on necessary truth could be the keystone of scientific methodology. If we could find such methods, then we would not need any external basis for methodology. But in order to accept this distinction we have to accept the idea that there are methodological truths and that some of these truths are self-evident or self-justificatory, or in other terms categorical imperatives and rational principles. Yet, it seems to contradict the point made in I.3 quite directly.

Even if we change our mind on this point, and the philosophical significance of scientonomy would be greatly affected by such a *volte-face*, the next difficulty would lie in the identification of procedural methods. These methods seem then to be the most abstract ones, the ones stating the condition of possibility for any scientific inquiry. They can be distinguished from the substantive ones that are only particular applications of this method, based on the internal interpretation of the scientific discipline itself. We cannot choose not to use these methods because both the origin and the reason why we use them escape our sight. It follows that procedural methods are static: once adopted, they cannot be replaced. This seems to also imply that procedural methods are not only necessary but also necessarily known as we cannot be mistaken about their nature or their content. It may be so, but if it is the case, then we face another contradiction since TSC does seem to be equivalent to MTD, at least concerning those methods, and therefore ultimately normative. Furthermore, these methods are not accessible through the description of scientific change but deduced from abstract propositions, just like an *a priori* methodology.

The nonempty mosaic theorem (Barseghyan, 2015, p. 226) tries to solve this difficulty, but it rather illustrates it. The question is: where do our most fundamentals methods come from? The goal is to describe how the groundings are grounded. It is brilliantly formulated: we must have rational expectations for scientific change to be possible. But I think that we cannot just state those expectations or deduce them from a scientific value such as truth, validity, or coherence, we have to understand where those expectations really come from. In a way, we have to understand the origin of the scientific attitude. The formulation of the fundamental method is interesting (Barseghyan, 2015, p. 230):

We have come across that requirement on many occasions. I am referring to the most abstract requirement to accept only the best available theories. This basic requirement is the most abstract of all, for it does not presuppose any other methods or theories. It is not surprising given that this abstract method is only a restatement of the definition of acceptance: this abstract method basically says that a theory is acceptable when it is the best available description of its object, i.e. acceptable. But since this abstract requirement isn't based on any theories, it cannot become accepted; it must be built into any mosaic from the outset.

Is this idea of maximum expectation a solid starting point? I have three objections:

- It does not presuppose any prior theory, but it does presuppose something. Those theories are the best at doing what, exactly? If it is anything, then the statement seems tautological: we are looking for something because it is worth looking for. Besides, pursuit worthiness is not necessarily synonymous with scientificity (Sarwar & Fraser, 2018). So we must have some idea of what we are trying to do with those theories. Science is not totally cut off from our other motives. What makes us engage in this theoretical competition? Why can't we tolerate any theory anyone could imagine? Hasn't it been actually the case throughout history?

- It also does not account for the use of necessary truths as procedural methods. They are not analytically comprised in the fundamental method of accepting only the best theories, so they have to come from somewhere else or they must be accepted as the second step of any method, but they are not without presupposition anymore. Would it be possible, for example, to derive a set of logical axioms, or even the rule that the best scientific discourse is always the most logically sound one, from the principle of maximum expectation or the general idea of method? The ambiguity about the changeability of scientific method remains: one could say that we only deepen our understandings of an efficient but contingent way to think about the world.

- If the most fundamental principle of scientific change is a principle of selection via random variation. Wouldn't it amount to accepting an evolutionary law? Should we speak of the evolution of theories and methods through selection then? But then again, an evolutionary principle is not a source of logical necessity, so the idea of static procedural methods becomes problematic.

5. Outline of a Theory of Problems

At this point, we have reached the clearest form of a philosophical problem, i.e., a dilemma:

- If there is no unchanging scientific method, then there is no method that would be intrinsically scientific, and we would be incapable of justifying in an absolute manner the facts that we categorize some activity as science and that a scientific theory is better than an unscientific one.

- But if there is a fundamental principle of scientificity then we have to understand where it comes from and the answer cannot be found inside the process of scientific change itself.

Therefore, to bring closure to a TSC, we must be able to describe how a method becomes scientific, how the idea of science arises. Here, the evolutionary perspective of BH and CE comes useful. In both programs, understanding the foundations and the basic laws of any level of change requires to understand how it emerges from a more basic one. To put it simply, if there is change, there is a reason for that change, there is a positive imbalance and it can be traced back. That is why understanding change requires to think big, because change cannot be understood inside a close field of inquiry. According to cosmic evolutionists, we can formulate a unique theory which would describe the various scales and stages of cosmic history, from primal energy to atoms, from atoms to galaxies, from astronomical objects to stable chemical elements, from chemical compounds to unicellular organisms, from simple life-forms to intelligent individuals and finally to advanced civilization (Chaisson, 2006, p. 12).

Thus, as a theory of change, TSC could parallel CE and articulate different levels of epistemic change in an interdisciplinary approach. By doing so, we would not have to stop at a first principle: we could access to a genealogy of the scientific attitude. To understand the fundamental laws and conditions of possibility for scientific change, we would have to understand how humans came to organize the evolutive selection of theories. This is precisely the subject matter of the evolutionary epistemology program: describing "the evolution of method constitutes the transformation of individual subjective and intersubjective knowledge into the transsubjective objective knowledge of science" (Oeser, 2005, p. 299). Hence, evolutionary epistemologists aim to synthetize anthropological evolution and TSC: to understand how science came to be, we must understand why men came to change, and to understand why science has to change, we must understand its place in the general interaction between humans and their world.

It is right to say that providing MTD a basis, i.e., a basic definition of science, is the same thing as providing scientonomy with a starting point: it allows one to decide when we can start describing an intellectual behavior as scientific and when we cannot. But it is possible without postulating necessary methodological truths. It is sufficient, and far more coherent with scientonomy's principles and scope, to describe how science originated and where it came from. As Oeser indicates, the answer can be found at the junction of paleo-anthropology and history of science. To understand what we are doing when we do science, we must describe how we went from our capacity to perpetually adapt our individual behavior to our environment to the organization of a collective inquiry about seemingly abstract and useless preoccupations.

There are different schools of evolutionary epistemology, such as that of Wuketits and Oeser (Austrian heirs of Lorenz and Popper), Jean Piaget's genetic epistemology, and Stephen Toulmin's theory of human understanding. Yet, they all share a common trait with Chaisson's conception of change: it comes from an imbalance caused by new information, which is another name for a *problem*. If we design scientific methods, it is because we face problems we could not solve otherwise, just as every new form of order in a complex dynamical system is a way out of a difficult situation, i.e., from the instability created by complexity itself (Chaisson, 2001, pp. 52-54). Therefore, from an evolutionary point of view, the origin of the scientific stance must lie in problems which are not yet scientific.

In the second half of the twentieth century, both science (Laudan, 1977) and human intelligence (Newell, Shaw, & Simon, 1958) have been characterized in terms of problem-solving, giving birth to renewed perspective not only in epistemology and cognitive science, but also in the history of human intellectual practices (de Beaune, Coolidge, & Wynn, 2009). During the same period, the concept of problem itself has been deeply explored by philosophers (Nickles, 1978, 1981). My aim here is not to present or even sum up these major contributions to the understanding of human knowledge. I simply wish to outline the possibilities opened up by combining this conception with an evolutionary approach of the changeability of method: a theory of problems that parallels the theory of method and may help to clarify some of the ambiguities I pointed earlier.

In an evolutionary setup, problems are not just lexical puzzles. If science is a problem-solving activity insofar as all life is problem-solving, the kind of problem we are interested in are concrete situations where no pre-existing strategy for ensuring self-preservation or achieving satisfaction has been effective. These are situations of imbalance that demand creativity and compel to innovation (Popper, 2010, p. 63).

In our epistemic lives, problems may arise from the limitations of our individual cognitive apparatus, that cause errors and mistakes, but also from its success which unveils unforeseen possibilities. Our ability to evaluate and organize our practical interactions with the material world can be internalized and turned on itself (that is what Piaget calls *abstraction réfléchissante*, reflecting abstraction) so that we can imagine new ways of using what we know and new limits to overcome, without having to experience them. But not all problems are personal problems, they can also arise through communication: the harmonization of our way of thinking and acting with those who share our lives and constraints is one of the most fundamental problems of the human condition. Finally, epistemic problems can come from technical breakthrough from adjacent fields of activity. For example, the emergence of symbolic communication may play a decisive role in the birth of science by facilitating comparison between rival theories.

According to the evolutionary point of view, each level of complexity emerges from another simpler level, even the most basic method of science has to be conceived as an answer to a problem to be properly understood. Therefore, the statement "we want to accept only the best available theories" can be linked back to more fundamental problems which are not yet theoretical and do not even need to be verbalised, for example:

- How can I overpass my individual limitations?

- Can I improve the way I usually behave?

- How can I continue to share common goals with people when they are growing more and more different from me?

These problems' formulation echoes the three core mechanisms of cognitive ontogenesis discovered by Jean Piaget: assimilation, accommodation, and decentering (Piaget, 1950). He thought it was possible to derive every rational structure from basic adaptative mechanisms. Without necessarily committing to any strong recapitulation hypothesis, we can observe that they also seem to work in the case of scientific evolution. They would indeed provide a reason to formulate and compare theories, hence they could function as a condition of possibility for the emergence of the fundamental method of any scientific mosaic. Problems indicate the existence of a need for an explanation, they are the driving force of inquiry. Methods, then, can be conceived as means to answer this need. Our basic needs usually stay the same, but, as we get better at satisfying them, they tend to take a more refined form. Ancient moralists used to see a vicious circle here, and the source of the misery of the human condition, but it could also be conceived as the source of human theoretical creativity.

The traditional positivist narrative tells the story of how primal perplexity became philosophical meditation and, eventually scientific inquiry. There might be some truth there, but we do not need to rely on rational reconstructions anymore: we can trace the history of these problems and the different ways they have been solved through time. We can also observe how the solutions we came with at different time shaped our cultures, our epistemic practices, and our minds. It has been the focus of very active research fields in the last decades such as cognitive archeology, evolutionary psychology, and paleoanthropology (Liebenberg, 2012; Dunbar, 1996; Tomasello, 1999; Heyes & Huber, 2000; de Beaune, Coolidge, & Wynn, 2009; Malafouris, 2013). The use of problems allows an epistemological genealogy based on empirical data and thus can put a stop to the infinite regress. If the scientificity of a theory is defined by a method, whose scientificity is defined by a more abstract method whose scientificity is defined by a decision to pursue science, we do not know how we come to know what science is. But if we do know how practical problems gave rise to the rational norm which allows a collective effort to deepen our understanding of these problems and to conceive more efficient solutions, then we can use this method to establish how scientific problems came to be and how problems, methods and theories co-evolve.

Therefore, my first suggestion regarding problems is that they should be integrated into the nonempty mosaic theorem: a mosaic is not empty insofar as it contains a problem. Problems are general because they come from the encounter of our intelligence, which is structurally the same in every human, and the most basic constraints of sociability and action, which are also common to everyone. That is why our scientific efforts are almost always oriented in the same direction. But those problems are also what drives the dynamics of scientific change, they are not questions that can be answered once for all but constant challenges. So they can also explain why science has to change, why theories have to compete. Therefore, methods are neither baseless, as they can always be linked back to another level of norms; nor transcendent, as they never rely on the idea that we could know something without a world to know something about.

My second suggestion is that TSC needs to explain the process of problem evolution in order to account for the change of the most fundamental parts of scientific method. Significant first steps toward this goal have been made through the inclusion of questions into the ontology of scientific change as fundamental epistemic elements (Rawleigh, 2018; Barseghyan, 2018). Yet questions are still theoretical entities: they formulate problems, more or less adequately, more or less simply, and they rely on a pre-existing language which express a pre-existing ontology. Thus, a single problem can be formulated by different questions belonging to different scientific disciplines (Nickles, 1981) or even to different scientific mosaics. Problems, however, have concrete existence: they are situations resulting from long-term conditions of our environment and by adapting to these conditions we build long-term knowledge which makes observational knowledge of short-term events possible (Popper, 2010, p. 64). Therefore, the claim that there is no universal and eternal scientific method can be argued far more clearly by relying on problems: some traits of a domain of inquiry are meant to remain relatively unchanged because they belong to the long-term problem that generates the field (the relation between our senses and the material world, the necessity to measure the limits of our strength, the temporality of our lives, the difference between thinking and acting, the irreducible diversity and creativity of human thought). It is thanks to those problems, that once again arise from the confrontation between a living intelligent being and a material world, that we can differentiate scientific domains.

But it is also evident that those problems evolve through the cultural and technological transformations that shape our relation to the world. At the same time, the way we are able to formulate these problems become more precise and complex, more theoretical: the introduction of the notion of force in physics, of transformation in biology, of signal in psychology, of logical consequence etc. have turned naively empirical inquiries into systematic descriptions of the underlying mechanisms of nature. The coexistence of stability and change in scientific mosaics come from the fact that the problems constituting the different parts of these mosaics are constantly evolving. Methods produce theories which can both give an answer to a problem and introduce new versions of the problem, new questions. Problems are the source of the infinite fertility of scientific inquiry, which has to be counted for by a TSC. Scientific methods can be conceived as hypothetical imperatives always renewed by the evolution of the problems they aim to solve. Therefore, it may be interesting to give problems a place in the third law of scientific change which describes how theories and methods produce new methods. Ideally, every theory should be connected to a question it answers and a question it asks. Furthermore, it could help us describe the complexification of mosaics, as part of scientific change: the multiplication of problems causes the complexification and atomization of the scientific world, which is a very concrete and concerning phenomenon of our time.

My third and final suggestion concerning problems deals with a peculiar type of problems, philosophical problems: the mind/body problem, the definition of life, the hard problem of consciousness, the necessity of physical laws, etc. From our point of view, we can see that most classical philosophical problems are transdisciplinary problems. I think that taking those problems into account could enrich our understanding of what it means for the elements of a scientific mosaic to be compatible, as compatibility itself can be dynamic since relation between domains can change according to the type of answer we give to these problems. My final suggestion is that the study of interdisciplinary problems could help us understand how changes in our scientific expectation travels from one part of a mosaic to another. Biological evolution is a good example, the fact that biology became able to investigate the origin of its most general structures and mechanism has contaminated the whole contemporary scientific mosaic, as the existence of CE demonstrates. But the mathematization of physics could do as well. It is through those problems that a new conception of scientificity progressively redefines our way of doing science, since a new methodology always arises from a renewed understanding of one of those objects (mind, body, matter, thought, laws).

Thinking big means choosing not to leave these problems to philosophy alone but to face them with our scientific knowledge, not with the hope of solving them once for all, or dissolving them into easier questions, but using them to grasp the full meaning of a scientific mosaic at any given time. It is certainly an optimistic stance, but I think it carries a better understanding of what we are trying to do when we pursue science and philosophy and the relation between these two quests.

This sketch of a theory of problematic evolution certainly lacks a technical formulation, but it illustrates what the evolutionary perspective can provide to TSC: a genealogical method that focuses on explaining how something new can happen. Reciprocally, it summarizes the contribution TSC can make to CE: a precisely formulated theory of the circulation of concepts and methods between different fields of inquiry, which is the greatest epistemological challenge for a general theory of change.

6. Conclusion

To conclude, I would like to formulate some answers to the questions I asked at the beginning of this paper.

- Can the Theory of Scientific Change (TSC) be used to solve the problem of scientific progress?

  Yes, if TSC takes a lesson form evolutionary epistemology, it can dissolve the problem of scientific progress by abandoning the idea of transcendent norms or any fixed aim and embracing the much less problematic idea of scientific evolution.

- Are the laws of scientific change reducible to some sociological or psychological laws?

No, but, at the same time, TSC must not make the mistake of conceiving science as a purely intellectual or ideal process. To understand the initial conditions of scientific change and the dynamics of scientific method we must consider the evolution of our understanding of our own cognitive abilities and societal issues. We should not abstract the laws of scientific change from the constraints and expectations they answer to.

- What is the place of problems (issues, questions) in the scientific mosaic? What role do problems play in scientific change?

It is a central role. They set the stage for scientific change to occur. They explain the relative fragmentation or unity of a mosaic. They are the link between our diverse scientific inquiry, the evolution of science, evolution in general and the evolution of philosophical systems.

Bibliography

Barseghyan, H. (2015). *The Laws of Scientific Change*. Springer.

Barseghyan, H. (2018). Redrafting the Ontology of Scientific Change. *Scientonomy*, 2, 13-38.

de Beaune, S. A., Coolidge, F. L., & Wynn, T.G. (Eds.) (2009). *Cognitive Archaeology and Human Evolution*. Cambridge University Press.

Chaisson, E. (2001). *Cosmic Evolution: The Rise of Complexity in Nature*. Harvard University Press.

Chaisson, E. (2006). *Epic of Evolution: Seven Ages of the Cosmos*. Columbia University Press.

Chang, H. (2016). Pragmatic Realism. *Revista de Humanidades de Valparaíso*, 8, 107-122.

Chang, H. (2022). The Ontology of Scientific Practice. In this volume, 1-19.

Christian, D. (2004). *Maps of Time: An Introduction to Big History*. University of California Press.

Cloud, P. (1978). *Cosmos, Earth, and Man: A Short History of the Universe*. Yale University Press.

Dunbar, R. I. M. (1996). *Grooming, Gossip, and the Evolution of Language*. Harvard University Press.

Gamble, C., Gowlett, J., & Dunbar, R. I. M. (2018). *Thinking Big: How the Evolution of Social Life Shaped the Human Mind*. Thames & Hudson.

Heyes, C. M. & Huber, L. (Eds.) (2000). *The Evolution of Cognition*. MIT Press.

Laudan, L. (1977). *Progress and its Problems. Toward a Theory of Scientific Growth.* University of California Press.

Liebenberg, L. (2012). *The Art of Tracking, The Origin of Science.* New Africa Books.

Malafouris, L. (2013). *How Things Shape the Mind: A Theory of Material Engagement.* MIT Press.

Newell, A., Shaw, J. C., & Simon, H. A. (1958). Elements of a Theory of Human Problem Solving. *Psychological Review,* 65(3), 151-166.

Nickles, T. (1978). Scientific Problems and Constraints. *PSA: Proceedings of the Biennial Meeting of the Philosophy of Science Association,* 1978, 134-148.

Nickles, T. (1981). What Is a Problem that We May Solve It? *Synthese,* 47(1), 85-118.

Oeser, E. (2005). The Evolution of Scientific Method. In Wuketits & Antweiler (Eds.) (2005), 299-331.

Piaget, J. (1950). *Introduction à l'épistémologie génétique - Tome I : La pensée mathématique.* Presses Universitaires de France.

Popper, K. R. (2010). *All Life is Problem Solving.* Routledge.

Rawleigh, W. (2018). On the Status of Questions in the Ontology of Scientific Change. *Scientonomy,* 2, 1-12.

Sarwar, A. & Fraser, P. T. (2018). Scientificity and the Law of Theory Demarcation. *Scientonomy,* 2(1), 55-66.

Shaver, P. A. (2011). *Cosmic Heritage: Evolution from the Big Bang to Conscious Life.* Springer.

Tomasello, M. (1999). *The Cultural Origins of Human Cognition.* Harvard University Press.

Toulmin, S. & Goodfield, J. (1982). *The Discovery of Time.* University of Chicago Press.

Wuketits, F. M. & Antweiler, C. (Eds.) (2005). *Handbook of Evolution: The Evolution of Human Societies and Cultures.* Wiley.

Chapter 8

Scientonomy and the Sociotechnical Domain

Paul Patton

University of Toronto

Abstract: The *sociotechnical domain* is the realm of scientists, the communities and institutions they form, and the tools and instruments they use to create, disseminate, and preserve knowledge. This paper reviews current scientonomic theory concerning this domain. A core scientonomic concept is that of an *epistemic agent*. Generally, an *agent* is an entity capable of intentional action—action that has content or meaning due to its purposeful direction towards a goal. An epistemic agent is one whose actions are the taking of epistemic stances, such as acceptance or rejection, towards epistemic elements, like theories or questions. An epistemic agent must semantically understand the propositions in question, and their alternatives, and choose among them with reason, with the motive of acquiring knowledge. The most obvious example of an epistemic agent is an individual human being. Rejecting the network of practitioners view, current scientonomic theory argues that appropriately organized communities of scientists can also function as epistemic agents. Communal epistemic agents are of particular scientonomic importance. Whereas the methods of theory assessment of individual scientists can be idiosyncratic, scientonomic theory contends that the taking of epistemic stances by scientific communities is a lawful, rule-governed process. A second concept of central importance is that of an *epistemic tool*. A physical object or system is an epistemic tool for some epistemic agent if there is a procedure by which the tool can provide an acceptable source of knowledge under the method employed by that agent. The agent is then said to *rely* on the tool.

Keywords: sociotechnical domain; epistemic agent; communal epistemic agent; epistemic tool; social ontology; distributed cognition

***

1. Introduction

Scientific change is the process by which the theories accepted by some agent as the best available description of its object, and the questions taken as objects of inquiry, change over time. Scientonomy initially focused primarily on these elements and the process of change itself. However, science is an embodied activity practiced by communities of human beings. They rely on simple and sophisticated tools to conduct research. A number of scientonomic works have begun to explore the entities and relations of what I will here call the sociotechnical domain; the domain of the agents of scientific change and their tools. These works draw ideas from earlier theories of scientific change (Barseghyan, 2015), social ontology (Loiselle, 2017; Overgaard, 2017, 2019; Overgaard & Loiselle, 2016), and the philosophy of cognitive science and biology (Patton, 2019) as sources of inspiration in formulating an ontology of the sociotechnical domain. The body of scientonomic literature reviewed in this chapter offers an alternative to the networks of practitioners view advocated by Bruno Latour and some others (Latour, 1987, 2005; Latour & Woolgar, 1986). It incorporates a robust view of the role of communities, instruments and tools in science, as well as the concept of distributed cognition that has emerged and gained currency in cognitive science and associated philosophy. In a distributed cognitive system, cognitive processes extend beyond the minds of individual agents to encompass tools and instruments, and other agents (Clark, 2001, 2007, 2008, 2010; Clark & Chalmers, 1998; Giere, 2002, 2003, 2004, 2007; Giere & Moffatt, 2003; Palermos & Pritchard, 2016; Palermos, 2011).

2. The Epistemic and Sociotechnical Domains

Scientonomy is a descriptive field that posits that the process of scientific change observes general principles, which it seeks to uncover (Barseghyan, 2015).[1] As a human activity, the production of knowledge can be understood in terms of two closely interrelated domains, which we will here call the *epistemic domain* and the *sociotechnical domain*. Loosely speaking, the epistemic domain is the realm of scientific ideas and the sociotechnical domain is the realm of scientists, the social communities and institutions they form, and the tools and instruments they use in the course of creating, disseminating, and preserving knowledge. The epistemic domain has so far been the principal focus of scientonomy. We will begin by briefly reviewing its main features as

---

[1] https://www.scientowiki.com/Scientonomy_(Barseghyan-2015)

they are currently envisioned by scientonomic theory, before turning our attention to our primary topic; the sociotechnical domain.

The *epistemic domain* can be defined as the set of all epistemic elements. Two sorts of epistemic elements are currently accepted within scientonomy; *questions* and *theories*. A question is a topic of inquiry (Rawleigh, 2018)[2] and a theory is a set of propositions (Barseghyan, 2018; Sebastian, 2016).[3] There are three sorts of theories, *definitions*, *descriptive theories*, and *normative theories*. A definition states the meaning of a term (Barseghyan, 2018).[4] A descriptive theory is a set of propositions that attempts to describe something (Sebastian, 2016),[5] and a normative theory is a set of propositions that attempts to prescribe something (Sebastian, 2016).[6] A *method* is an especially notable type of normative theory. It is a set of requirements for employment in theory assessment (Barseghyan, 2018).[7] Theories are related to questions and to one another. A theory is an answer to a question, and a question can presuppose a theory (Rawleigh, 2018). For example, the question 'What is the mass of the electron?' presupposes theories that electrons exist and have mass. The accepted answer to this question is the theory that states that the mass of an electron is 9.1×10^{-31} kilograms (NIST, 2018).[8]

The *sociotechnical domain* includes both communities of human beings and the instruments and tools they use to generate and transmit knowledge. From Plato's Academy in Ancient Greece, to the Intergovernmental Panel on Climate Change today, social cooperation has played a central role in the creation of scientific knowledge (Barseghyan, 2015, pp. 43-52; Overgaard, 2017, 2019; Patton, 2019). Tools and artifacts used for observation, measurement, recording, computation, as an aid to reasoning, and as a means to preserve and disseminate symbolically expressed epistemic elements likewise play a central role in the process of scientific change. Tycho Brahe's careful measurements with quadrant and sextants, Galileo's use of the telescope, and the recording of their observations with quill and paper were clearly important to the formulation and acceptance of heliocentric astronomy in the sixteenth and seventeenth centuries. Experiments with giant particle accelerators, calculation and reasoning with pencil and paper, blackboard and chalk, and powerful computers

---

[2] https://www.scientowiki.com/Question_(Rawleigh-2018)

[3] https://www.scientowiki.com/Theory_(Sebastien-2016)

[4] https://www.scientowiki.com/Definition_(Barseghyan-2018)

[5] https://www.scientowiki.com/Descriptive_Theory_(Sebastien-2016)

[6] https://www.scientowiki.com/Descriptive_Theory_(Sebastien-2016)

[7] https://www.scientowiki.com/Method_(Barseghyan-2018)

[8] Recently Kye Palider (2019) has proposed another sort of relationship connecting theories with other theories – that of *reason*.

and the dissemination of ideas through printed journals available in libraries were equally essential to the formulation and acceptance of the standard model of particle physics in the twentieth century. Even in the formal sciences, physical tools are essential to all but the most rudimentary mathematical and logical computations because of the limitations of human memory and computing ability (Rumelhart & McClelland, 1986, pp. 44-48; Dehaene, 2011). In what follows, I will review past work towards the development of a scientonomic theory of this sociotechnical domain.

3. Entities and Relations in the Sociotechnical Domain

In order to incorporate the sociotechnical domain of scientists and their tools and instruments into scientonomic theory, we need a guiding ontology that identifies the relevant entities and their relations to one another. Barseghyan (2015, pp. 48-52) considered two levels at which we might look for such sociotechnical entities and relations – the level of the beliefs of the individual scientist and the level of the scientific community and its mosaic of accepted theories and employed methods.

According to Overgaard (2019, pp. 12-64), Barseghyan's levels correspond to two different approaches that have guided research in the philosophy, history, and sociology of science. He calls them the conceptual frameworks camp and the networks of practitioners camp. The conceptual frameworks camp consists of those scholars who regard intellectually and culturally unified communities of researchers as the bearers of knowledge and the units of analysis we must study to understand the process of scientific change. Such communities might include the scientific community as a whole, a particular disciplinary community, or an individual research lab. Many past theories of society, including past theories of scientific change, like those of Kuhn and Lakatos, have been grounded in the conceptual frameworks view (Overgaard, 2019, pp. 12-40).

Critics of the conceptual frameworks camp contend that, when scrutinized carefully, social communities dissolve into a complex hash of unique and distinctive individuals and their relationships to one another, with no clear boundaries to this network of relationships. The concept suffers, it is claimed, from the same problems as essentialist classifications of organisms in modern evolutionary biology (Overgaard, 2019, pp. 32-39; Mayr, 2006; Sober, 2006), which flounder over the unique and distinctive combinations of traits produced by sexual reproduction and the independent inheritance of genes. Rather than being ontologically real features of the social world, communities are artificial illusions foisted onto an unruly world by social theorists.

Having decided that communities do not exist, the networks of practitioners camp focuses on individual scientists and the networks of material and social interactions in which they engage. The most noted such approach is Latour's actor-network theory (Overgaard, 2019, pp. 41-47; Latour, 1987, 2005). Latour's theory supposes an indefinitely large network of actors in which both human beings and instruments and tools play symmetrical roles and possess agency. The claim that tools and instruments possess agency is Latour's way of capturing the causal role that the structure of the natural world plays in science. The network of practitioners framework suffers its own set of ontological and other problems. Overgaard (2019, p. 63) notes that networks incorporating social and natural entities on an equal footing blur the distinction between the social and the natural world. Giere (Giere, 1992; Giere & Moffatt, 2003) has also pointed out that the theory ignores relevant and important ideas deriving from cognitive science. Rejecting what they suppose is a dubious focus on communities and abstract theories, proponents of the network of practitioners approach seek to ground the study of science empirically by the direct study of scientific practice in the laboratory, shifting the focus away from scientific theories. Overgaard (2019, pp. 61-64) takes this approach to be the dominant one in current science studies.

From its beginning, scientonomy, like earlier general theories of scientific change, has been aligned with the conceptual frameworks camp, taking the concept of communities, and what we have called epistemic elements to be centrally important to the process of scientific change (Barseghyan, 2015, pp. 43-52). Barseghyan (2015, p. 43) wrote that "when we speak of some transformation in science, we don't mean that this or that great scientist has changed her mind and decided to accept a new theory or employ a new method, but that the scientific community *as a whole* has rejected some elements of the mosaic and replaced them with new elements". He further argued that the past failure to find a lawful process of scientific change has been due to the historian's tendency to conflate the individual with the social, and to focus on the idiosyncrasies of prominent individual scientists rather than on the behavior of scientific communities as a whole (Barseghyan, 2015, pp. 45-47). Further, he supposed that the behavior of the community is neither determined by elite individual scientists, nor is it the simple summation of the individual views of all the members of the community. He supposed instead that the relationship between the individual and community involved complex social dynamics (Barseghyan, 2015, pp. 48-51).

Subsequent scientonomic work concerning the sociotechnical level has thus focused on justifying the existence of communities as distinct and ontologically real social entities amenable to scientific study. Such a justification was provided by the work of Overgaard (Loiselle, 2017; Overgaard,

2017, 2019; Overgaard & Loiselle, 2016), who drew on the findings of social ontology (Lawson, 2014; Searle, 2006; Tollefsen, 2014; Tuomela, 2002) to argue that what he called *epistemic communities* were both ontologically real and methodologically accessible for study. Overgaard's work was further refined and elaborated by Patton (2019), who drew inspiration primarily from work in the philosophy of cognitive science and biology (Clark, 2001, 2007, 2008, 2010; Clark & Chalmers, 1998; Giere, 2002, 2004, 2007, 2010; Palermos & Pritchard, 2016; Palermos, 2011; Palermos, 2016; Theiner, 2014; Theiner, Allen, & Goldstone, 2010; Theiner & O'Connor, 2010; Wimsatt, 2006, 2007) to formulate the concept of an *epistemic agent*, that may be either individual or communal. I sought to explicate the nature of the emergent relationship between individual and communal epistemic agents. I also explicated the role of scientific tools and instruments as *epistemic tools* (Patton, 2019). Following Clark, Giere and others, I maintained the central importance of distributed cognition in understanding this role. In this chapter, I will review this past work and survey our current understanding of the entities and relations of the sociotechnical domain.

4. Agency, Intentionality, and Epistemic agency

To understand the role of individual human beings, and communities of human beings in the production of knowledge, and to discern whether and how that role is distinct from that played by tools and instruments, we need a definition of *epistemic agent*; the actor in the process of scientific change. We begin with the more general concept of an *agent*. An agent has typically been defined as an entity capable of *intentional action* (Schlosser, 2015). To understand what is meant by intentional action, we must first consider the more general concept of *intentionality*.

Intentionality is the property possessed by representational states that have content or meaning. Such states are *about* something. They refer to an object. Mental states, like perceptual states or belief states, are the classic examples of states that possess intentionality (Dretske, 1981; Jacquette, 2006; Neander, 2012, 2017, pp. 63-82; Dennett, 1971; Millikan, 1984, pp. 85-94; Overgaard, 2019, pp. 65-70). Many philosophers no longer regard them as the only intentional states. Symbols with semantic content, in a medium external to the body, like spoken words in air, text on paper, instrument readings, computer displays, and software code are taken to exhibit intentionality as well (Dretske, 1981; Jacob, 2019). Proponents of the extended mind thesis argue that such symbols, though external to the body, should not be regarded as external to the mind (Clark, 2008, 2010; Clark & Chalmers, 1998). Leaving aside the question of whether or not such items are constituent parts of the mind, proponents of distributed cognition regard them as intentional bearers of cognitive content (Giere, 2002; Giere &

Moffatt, 2003; Rumelhart & McClelland, 1986, pp. 44-48). Other biological states, such as the arrangement of nucleotide base pairs in the genome of an organism, are also said to possess intentionality, since they perform the function of containing semantic information for the system of which they are a part (Dretske, 1981; Fitch, 2008; Godfrey-Smith, 2006; Godfrey-Smith & Sterelny, 2016). This broad understanding of intentionality will be of considerable importance when we turn to the discussion of epistemic tools. The suggested definitions are presented in Figure 8.1.

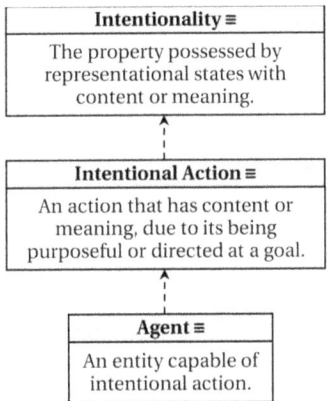

Figure 8.1: The taxonomy of 'intentionality', 'intentional action', and 'agent'

When we say that an agent's actions are intentional, what we mean is that they have content or meaning because they are purposeful, directed towards a goal, or performed for a reason. By some accounts, this is because they are caused by the agent's mental states. This has made the concept of agency problematic for naturalists, since, as traditionally understood, it seems to involve dualistic notions of mental causation that are irreconcilable with physical causation (Jacquette, 2006; Schlosser, 2015). But, a number of authors have proposed naturalistic accounts of agency in which intentional states supervene on appropriate physical states, or are simply a way of interpreting those states. On this account, mental or intentional causation is real, but as a species of physical causation, arising when a physical system has the appropriate special state of dynamical organization, such as that possessed by an organism and its brain in interaction with the world (Dennett, 1984, 1987, 1991, 2003; Fulda, 2016; Giere, 2004; Schlosser, 2015; Walsh, 2012, 2016; Walter, 2009, pp. 239-268; Thompson, 2007, pp. 37-65). More broadly, agency appears to be a universal organizational feature of living systems, evident in the flexible goal-directed behavior of bacteria and other single-celled organisms, as well as in the behavior of the cells that make up multicellular organisms (Walsh, 2015, pp. 208-229; Fitch, 2008; Fulda, 2016, 2017; Walsh, 2016). It has also been argued that a derived form of agency is present in goal-seeking engineered systems as well, with thermostats being a simple example (Dennett, 1987, pp. 37-42).

Barseghyan (2018) was the first to introduce the concept of an epistemic agent in scientonomy, and a definition and theory was formulated by myself (Patton, 2019). My definition of epistemic agent is grounded in the general definition of agent explained above. Under that definition, an agent can be seen as an entity capable of perceiving its environment and acting within it with a motive or in pursuit of a goal, choosing among multiple courses of action to best fulfill that goal. For an epistemic agent, the relevant environment consists of epistemic elements, which, as we have noted, include theories, questions, and methods. The epistemic actions of an epistemic agent are the taking of epistemic stances towards these elements, such as accepting or pursuing a theory, or employing a method. To qualify as an epistemic agent, the agent's goal in doing so must be to acquire knowledge of the world (Figure 8.2).

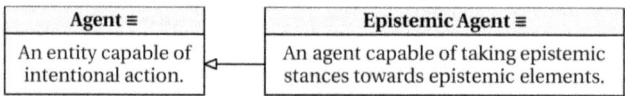

Agent ≡	Epistemic Agent ≡
An entity capable of intentional action.	An agent capable of taking epistemic stances towards epistemic elements.

Figure 8.2: The definitions of 'agent' and 'epistemic agent'

Thus, a random number generator, selecting among various versions of string theory proposed by physicists is not an epistemic agent. To constitute an exercise of epistemic agency, an agent's taking of an epistemic stance must be an intentional action, and for this to be the case, two conditions must be met: 1) the agent must have a semantic understanding of the propositions that constitute the epistemic element in question and its available alternatives, and 2) the agent must be able to choose among the available alternatives with reason, and for the motive of acquiring knowledge. In scientonomic terms, the normative epistemic strategy which an agent deploys in assessing a theory is the agent's *method* (Barseghyan, 2018).[9] When fabricating and planting the fossils, the perpetrator of the Piltdown Man hoax, most likely Charles Dawson (Donovan, 2016), was not acting as an epistemic agent since he almost certainly did not accept the theory that forgery was a valid route to knowledge of human evolution. Epistemic agency involves the employment of the norms of epistemic honesty one accepts (Goldberg, 2016). When astronomer Arthur Eddington made his careful and meticulous observations of stars near the sun in the sky during a solar eclipse, he was acting as an epistemic agent. This is

---

[9] https://www.scientowiki.com/Method_(Barseghyan-2018)

because, under the method he employed, Einstein's theory of general relativity was more likely to be true if its unexpected novel prediction that the gravitational field of the sun bends starlight were correct, and because Eddington employed the rigorous standards of observation, measurement, and epistemic honesty that he accepted as necessary for astronomical observation (Isaacson, 2007, pp. 255-262).

Who or what can be an epistemic agent? Scientonomy takes epistemic elements to be propositional. It has, so far at least, also taken propositions to be sentential; that is, they are expressible by a sentence or sentences of a natural language, mathematics, or logic. Thus, non-linguistic animals and pre-linguistic human infants can be ruled out as epistemic agents, since they lack the requisite ability to semantically understand sentential propositions. This leaves the most obvious example of an epistemic agent as the typical individual human being. Such an individual can, given appropriate education and training, understand the propositions that constitute an epistemic element, and its alternatives. They can choose among these alternatives with reason, with the goal of acquiring knowledge of the world. It should be evident that individuals may vary, one from another, in the degree to which they satisfy the definition of an epistemic agent. Their degree of semantic understanding of the epistemic elements in question, for example, may vary with experience, education, and professional training. The sincerity of their commitment to the goal of acquiring knowledge of the world may also vary, as indicated by their adherence to norms associated with epistemic honesty.

5. Communities as Epistemic Agents

Whenever epistemic elements are explicitly stated as sentential propositions, they can be shared with other epistemic agents. While typical individual human beings clearly do satisfy our definition of epistemic agent, the point of central importance for scientonomy is whether or not communities that include multiple individuals can do so. In order to claim that communal epistemic agents are ontologically real, we must show that they satisfy the conditions for epistemic agency in their own right, as distinct from the individual epistemic agents which they include as constituent parts. If they do really exist, then any account of scientific change that left them out would be seriously deficient. Further, they, and the circumstances which bring them about, must be amenable to empirical study through historical or sociological research. I will argue here that they do exist and are amenable to empirical study, drawing primarily on the scientonomic work of Overgaard (2017, 2019). The idea of social communities as ontological elements and as agents has been developed in depth by social ontologists (Lawson, 2014; Searle, 1995, 2006; Tollefsen, 2014; Tuomela, 2002), and has been applied to the concept of epistemic communities by a number of authors (Overgaard, 2017, 2019; Palermos & Pritchard, 2016; Palermos, 2016; Theiner & O'Connor, 2010; Tollefsen, 2004, 2006).

The taxonomy diagram in Figure 8.3 presents definitions and relationships associated with epistemic communities as formulated by Overgaard (2017, 2019).

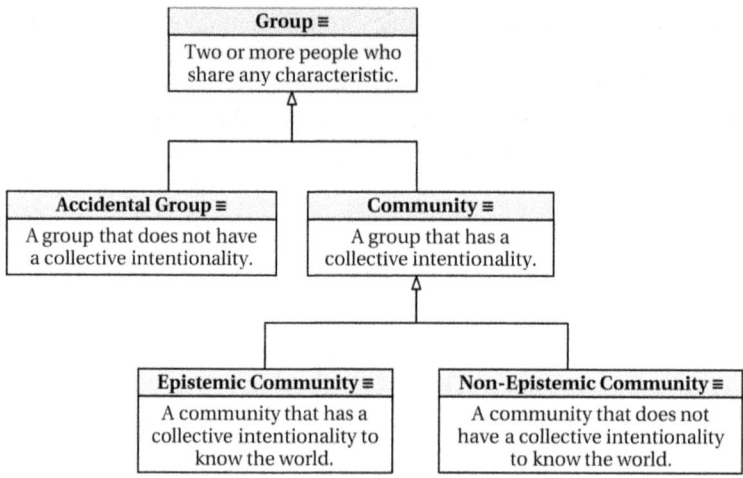

Figure 8.3: The taxonomy of 'group', 'community', and 'epistemic community'

A *group* consists of two or more individuals sharing any characteristic. The set of people with blue eyes who are fans of the Green Bay Packers football team is an example. Groups may be defined, essentially arbitrarily, based on any collection of traits. *Accidental groups* are those groups that just happen to contingently share some collection of traits. Members of such a group need not necessarily be agents. Social ontologists distinguish a *community* as a very special sort of group consisting of agents that share a *collective intentionality*, that is, that are organized to jointly pursue a collective goal or purpose (Overgaard, 2017, 2019).

Collective intentionality is not simply shared intentionality. A group of parents that share an intention to raise their children do not share a collective intentionality because they each direct similar intentional actions separately towards their own children. A football team or a symphony orchestra does share a collective intentionality because the intentional actions of individuals make distinctive coordinated contributions towards a singular collective goal- -such as winning the game or playing the symphony. The sharing of a collective intentionality is not merely an incidental or accidental feature of membership in a community; it is the *essential* feature of membership, by which members can be distinguished from non-members.

Overgaard (2017) defines an epistemic community as one that has a collective intentionality to know the world. More specifically, an epistemic community may seek to answer a particular question, or set of questions that have been accepted as legitimate topics of inquiry (Rawleigh, 2018). Those questions might, for example, be those delineating a particular scientific discipline (Patton & Al-Zayadi, 2021). Members of that particular epistemic community can be identified by their coordinated efforts with other members of that community to answer that question or those questions that are the particular object of that community's collective intentionality. The members of a research group might coordinate their actions to perform an experiment aimed at answering a particular question, with each making their own distinctive contribution towards its success. The outcome might be a single co-authored published paper containing a theory which answers the question at issue, and is accepted by the entire team.

Just as individual agents can be constituent parts of an epistemic community if they perform a distinctive coordinated role within the collective intentionality of that community, so can other epistemic communities (Overgaard, 2017). Thus the scientific community of the modern world shares a collective intentionality to know the world by answering all those questions it accepts as legitimate topics of inquiry under its demarcation criteria. Particular disciplinary communities play distinctive specialized roles in fulfilling this larger goal (Patton & Al-Zayadi, 2021). The community of physicists deals with those questions generic to the behavior of matter and energy, the astronomical community answers those questions dealing specifically with celestial objects, the geological community with those questions dealing specifically with the constituent parts of the Earth and other such planetary bodies, the biological community those questions distinctive to living systems, and the psychological community those questions specific to the mental processes and behavior of living systems. Such a hierarchy of communities and subcommunities continues to the level of individual research laboratories and their members. A hallmark feature of collective intentionality is a *division of epistemic labor* in which different epistemic agents perform distinctive specialized roles towards the fulfillment of a goal that no one person could possibly possess the skills, education, and training to accomplish by themselves.

For an epistemic community to qualify as an epistemic agent in the sense defined above, it must be capable of taking stances towards epistemic elements. Those stances must distinctively belong to the communal agent itself, rather than to its constituent agents, taken separately. There are good reasons for supposing that an epistemic community can possess some

properties that belong distinctively to it, rather than to its constituent
individual agents taken separately. Systems with multiple interacting parts, if
those parts are appropriately organized in relation to one another, exhibit
emergent properties that are the product of that organization rather than of any
of its constituent parts, taken separately (Bedau, 1997; Kim, 1999; O'Connor &
Yu Wong, 2015; Wimsatt, 2006, 2007, pp. 274-312). Such properties belong to the
system as a whole rather than to any of its parts.

Wimsatt (2006, 2007, pp. 274-312) defined the emergent properties of a
system as those that depend on the way its parts are organized. An *aggregate
system* is one whose parts do not bear an organized relationship to one another.
The parts all play similar roles and can be interchanged or rearranged without
any consequence. The properties of the whole are an additive, statistical
consequence of those of its parts. There are no emergent properties. A jumbled
pile of mechanical parts is an example of an aggregate system. Its properties,
like its mass or its volume, are just the sum of the masses and volumes of its
parts.

A *composed system,* on the other hand, is one that possesses emergent
properties due to the way in which its parts are organized in relation to one
another. A clock assembled by arranging mechanical parts in the proper causal
relationship to one another is an example of a composed system. The clock's
ability to indicate the time of day is an emergent property, because no part of
the clock possesses that ability on its own. The parts are organized so that there
is a division of labor among them, and each plays its own distinctive role in the
production of the emergent property. As we have seen, collective intentionality
requires that a community be organized in such a way that each individual
agent plays a distinctive role in the fulfillment of the community's shared goal.
Thus, communities, including epistemic communities, seem likely to possess
emergent properties belonging specifically to them (List & Pettit, 2006, 2011;
Overgaard, 2019; Palermos & Pritchard, 2016; Palermos, 2016; Patton, 2019;
Theiner, Allen, & Goldstone, 2010; Theiner & O'Connor, 2010, Wimsatt, 2006,
2007).

To satisfy our definition of an epistemic agent, an epistemic community must
be organized such that its decision-making processes lead to epistemic stances
that are emergent properties of the community as a whole, rather than the
simple aggregate of the decisions of its individual members. Given the
properties of epistemic communities as we have outlined them, this seems
quite likely to be the case. The individual agents that make up an epistemic
community can, by definition, semantically understand epistemic elements.
Because they will each have at least somewhat different areas of expertise, they
will each bring a different area of semantic understanding to the decision-
making process of the community as a whole. Different agents, for example,

will be better equipped to assess different premises of an argument. Interactions among such diverse individuals with different areas of expertise would seem to ensure that the views of individual community members are influenced by others, leading the community to take epistemic stances that are distinct from those the same individuals might take if left to their own devices. We will discuss below the concept of authority delegation, in which different epistemic agents are recognized as possessing different areas of expertise in a division of cognitive labor.

In an analysis of legal decision-making, Tollefsen (2004) outlines a simple scenario in which the decisions of a committee might be viewed as emergent and belonging to the committee rather than to its individual agents. The members of a committee are asked to assess each of the premises of an argument separately. In such a case, the conclusion reached by the committee, though following logically from the premises, might not be one that any individual would agree with. Given that different members of an epistemic community will bring different levels of expertise to different premises of an argument, and will be influenced by an awareness of the expertise of others, such scenarios seem an almost inevitable feature of such communities.

Tollefsen's investigations of legal decision-making demonstrate that the weakly emergent properties posited here are in no sense mysterious. Their occurrence in epistemic communities can be investigated empirically in much the same way that Tollefsen did for legal communities. Regardless of the details of the social mechanisms by which they arise, communal epistemic agents are a plausible foundation for scientonomy (Barseghyan, 2015, pp. 48-51; Overgaard, 2019; Patton, 2019, see Figure 8.4).

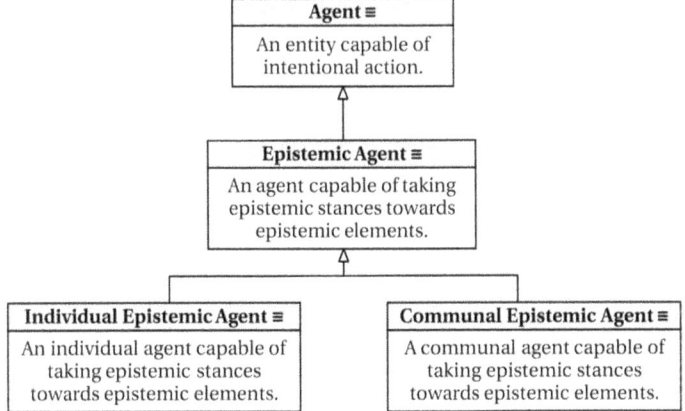

Figure 8.4: The taxonomy of agents

6. The Distinctive Importance of Communal Epistemic Agents

In order to grasp the distinctive importance of communal epistemic agents to scientonomy, it is helpful to consider both their similarities to individual epistemic agents and their fundamental differences, relying, in part, on some ideas from cognitive science. There are some important ways in which the two kinds of epistemic agents bear a closer resemblance to one another than one might, at first, suppose. Fitch (2008) has argued that the living cells that make up individual epistemic agents themselves exhibit a rudimentary form of intentionality and agency, which he calls "nano-intentionality". This nano-intentionality is manifested in the ability of cells to rearrange their own structure in response to damage, nutrient distribution, and other factors of their environment. Fitch regards cellular nano-intentionality as the foundation for the intentionality exhibited by multicellular organisms, including the epistemic agency of human beings. The functionalist view of mind sees cognition as an emergent consequence of the organized causal interaction among elements that are not themselves cognitive (Bechtel, 1988, 2008; Clark, 2008; Levin, 2018). The epistemic agency exhibited by individual epistemic agents is thus a species of collective intentionality, emerging from the organized interaction of vast numbers of individual elements, each possessing only nano-intentionality and together constituting a composed system which exhibits epistemic agency as an emergent property. Investigating the dynamical mechanisms by which epistemic agency arises from the interactions of simpler elements is the subject matter of cognitive neuroscience (Bechtel, 2008; Clark, 2015).

While individual and communal epistemic agents may resemble one another in that both possess cognitive processes emerging from their composed organization, there are few good reasons to suppose they are similar in many other respects. For example, social scientist Karin Knorr-Cetina (2009) has proposed that communal epistemic agents are similar to individual epistemic agents in possessing their own consciousness. This doesn't seem particularly likely given the profound disanalogies between these two sorts of cognitive systems. It also is unclear how one might recognize consciousness in such an unfamiliar form. While the similarities of the two kinds of systems should be noted, their dissimilarities are also profound, and as we will see, critically important to a scientonomic theory of the sociotechnical domain.

An individual epistemic agent consists of a vast number of interacting living cells, in particular, the 86 billion nerve cells, or neurons, of the human brain (Azevedo et al., 2009; Herculano-Houzel, 2009), each of which communicates with as many as ten thousand others by way of patterned discharges of neurotransmitter substance, a relatively simple sub-symbolic vehicle of cognitive content (Bechtel, 2008; Dretske, 1981). By comparison, communal epistemic agents consist of vastly smaller numbers of individual elements. The largest such community known is the global scientific community of the modern world, which is estimated to consist of 7.8 million individual researchers (UNESCO, 2013), a number four orders of magnitude smaller than the number of neurons in the brain of a human individual. The most salient differences, however, concern the nature of the interactions between the elements.

Cognitive scientists once supposed that the inner workings of the individual mind were much like those of the outer world of symbols and formal, rule-based logic that humans have communally fashioned. Cognition was taken to be the inner manipulation of symbols in accordance with rules, as in formal logic, or the function of a digital computer (Fodor, 1975; Newell & Simon, 1976; Putnam, 1960; Turing, 1950). This view of the mind has been rejected. Beginning in the 1980's cognitive scientists began to attend to the architecture of the brain and to study the behavior of simulated networks of neurons (Bechtel & Abrahamsen, 2002; Churchland, 1989; Churchland & Sejnowski, 1992; Dayan & Abbott, 2001; Rumelhart & McClelland (Eds.), 1986). By the end of the century, cognitive neuroscientists saw the mind/brain as a dynamical system whose neural parts were engaged in nested loops of interaction among themselves and with the body and the world (Beer, 2000; Clark, 2015; Friston, 2003; Friston et al., 2017; Hohwy, 2013; Kelso, 1997; Rabinovich et al., 2006; Varela, Thompson, & Rosch, 1997). The partially analog, dynamical and sub-symbolic neural processing which forms the inner workings of individual agents mediates their sensorimotor interactions with the world and their inner cognitive workings, as well as their interactions with one another that do not involve symbolic language.

The exchange which takes place between the individual epistemic agents that make up a communal epistemic agent are of a fundamentally different nature than those occurring between the constituent neural and cellular parts of an individual epistemic agent that has been less frequently recognized. Scientonomy defines epistemic elements as propositional and stateable in sentences of natural language, logic, or mathematics, or as graphical diagrams. These forms of expression, as we have seen, appear not to reflect the inner workings of the mind, but rather the manner in which the outcome of those workings is expressed publicly. The emergent cognitive decision-making processes of communal epistemic agents arise in a very different way than do the inner workings of the individual mind. They emerge by the exchange of symbolic epistemic elements among limited numbers of individual epistemic agents assisted in their decision-making by propositional reason. The concept of epistemic communities as distributed cognitive systems has been explored by a number of researchers, including, notably, Fleck (Fleck, 2012, 1986), Giere

(Giere, 2002, 2004, 2007; Giere & Moffatt, 2003), Theiner (Theiner, 2011, 2014; Theiner, Allen, & Goldstone, 2010; Theiner & O'Connor, 2010), Palermos (Palermos & Pritchard, 2016; Palermos, 2011, 2016), and Patton (Patton, 2019).

From what we have seen about the distinctive cognitive properties of individual and communal epistemic agents, it would not be at all surprising to find that the epistemic stances of communal epistemic agents have new properties that are not present in those of individual epistemic agents. There are good arguments that such distinctive properties do exist. Longino (Longino, 1990, 1996, 2019) has argued that when communities have normatively appropriate structures, critical interactions between individuals holding different points of view and influenced by differing contextual biases mitigate the influences of their individual subjective preferences. This allows communities to obtain a level of objectivity in their taking of epistemic stances that is seldom possible for individual epistemic agents.

Similarly, Barseghyan (2015, pp. 43-52) has noted that the methods of theory assessment employed by individual scientists are often idiosyncratic, and attempts to identify a lawful process of scientific change through the study of individual scientists have not proved successful. Paul Feyerabend's conclusion that scientific change is not a rule-based process, for example, was based largely on his studies of Galileo's personal epistemic practices. He suggests that a lawful process of scientific change arises from the non-aggregative properties of communities of the sort developed in much greater detail here, and elsewhere in the literature of group cognition as noted above. Current scientonomic theory posits that the taking of epistemic stances by communal epistemic agents is a lawful, rule-governed process, and a number of laws of scientific change have been identified by Barseghyan (2015, pp. 123-243) and other scientonomic authors (Fraser & Sarwar, 2018; Patton, Overgaard, & Barseghyan, 2017; Sarwar & Fraser, 2018). These laws, described by scientonomy, can be understood as implicit norms of rationality employed universally by communal epistemic agents. It should be stressed, however, that the goal of scientonomy is not to prescribe epistemic norms, but rather to empirically describe the norms actually employed by communal epistemic agents. The posited laws are amenable to empirical testing against historical evidence. We envision such testing as a central activity of what has been called observational scientonomy (Barseghyan, 2018).

7. Authority Delegation

We have already developed the idea of communal epistemic agents as composed systems in which different constituent agents play distinctive roles in a division of epistemic labor. The concept of authority delegation was formulated by Overgaard and Loiselle (Overgaard & Loiselle, 2016; Loiselle, 2017) to describe the relationship of interdependence between such epistemic agents having different areas of expertise. They supposed that this relationship was applicable to both communal and individual epistemic agents (Figure 8.5).

Authority Delegation ≡
Epistemic agent A is said to be delegating authority over question *x* to epistemic agent B *iff* (1) agent A accepts that agent B is an expert on question *x* and (2) agent A will accept a theory answering question *x* if agent B says so.

Figure 8.5: The definition of 'authority delegation'

As mentioned earlier, epistemic agency can vary as a matter of degree. Agents can vary in the degree to which they semantically understand the theory being assessed and its alternatives. They can also vary in their devotion to the goal of pursuing knowledge, as reflected in adherence to norms associated with epistemic honesty (Machado-Marques & Patton, 2021). Authority delegation reflects an assessment of such matters. The concept is reducible to theories and the stances of agents towards them. Overgaard and Loiselle used examples drawn from art-related agents (Overgaard & Loiselle, 2016; Loiselle, 2017), as indicated in Figure 8.6.

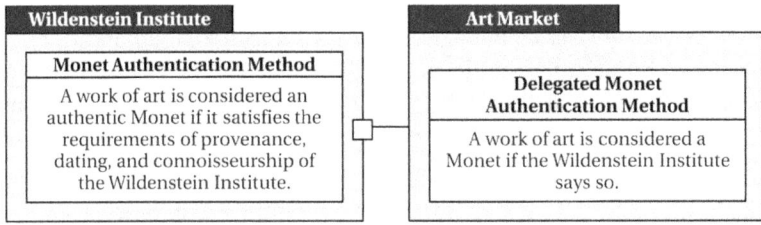

Figure 8.6: A diagram showing a case of authority delegation

The diagram indicates the employed method of the art market community and that of the Wildenstein Institute. The square symbol represents authority delegation. Loiselle noted that the art market community – the community responsible for buying and selling works of art – accepts the theory that the Wildenstein Institute is an expert on determining the authenticity of Monet paintings. Delegating to the Wildenstein Institute is thus its employed method, deduced from that theory, of accessing the theory that some particular painting was a work of Monet. The Wildenstein Institute's authority rests on its ability to justify its own employed methods for assessing the authenticity of a Monet painting, and its application in individual cases, in terms of the accepted theories from which it was deduced. Its authority also rests on the acceptance of the theory that its assessments are in accord with norms associated with epistemic honesty, and will not be swayed by non-epistemic factors, such as the offer of a bribe.

In populations of living organisms, traits are inherited in an independent and largely uncorrelated fashion, making each a unique individual and defeating all attempts to forge essentialist biological categories based on constellations of essentially shared traits. Differences in biology, environment, and upbringing likewise make each human being a unique individual, and make social scientists understandably leery of essentialist categories. But epistemic communities are not the product of undirected biological or social processes. They are intelligently designed social artifacts assembled for the purpose of pursuing knowledge. The assessment of theories regarding authority delegation by epistemic agents who semantically understand the theories they are assessing is one part of this design process. As in the design of a complex technological system, there are many designers and their actions may not be entirely coordinated with one another, but there is nonetheless a design process. Epistemic communities can thus be successfully identified as such based on the shared essential trait of collective intentionality. Note that not all members of such a community need to be aware of this collective intentionality in order to participate in it through their epistemic actions. In order to act as intelligent designers of such a community, however, it must be the case that at least some of its individual members harbor at least a partial semantic understanding of their creation. It is entirely possible that different designers of such a community understand different specialized aspects of its organization, and that the design of the community is itself an exercise of communal epistemic agency.

Since historical records reflect the designers' understanding of their own creation, the empirical study of epistemic communities does not seem to involve special difficulties. Membership in a community can be recognized by such things as the garnering of a faculty appointment in a particular discipline, membership in professional societies, and attendance of conferences. The professionalization of science is a phenomenon that began in the eighteenth and nineteenth centuries. The nature of communal epistemic agents prior to that time is an important matter for historical inquiry. Community acceptance of an epistemic element can be assessed, for modern communities at least, by noting the contents of textbooks, encyclopedias, and university curricula. Membership in a research group can typically be determined by co-authorship of published papers.

8. What are Epistemic Tools?

We have so far reviewed the role of epistemic agents in current scientonomic theory. I have argued that communal epistemic agents exhibit a kind of emergent distributed cognition that is distinct from the cognitive activities of their constituent individual agents taken separately. This is because communal epistemic agents are composed systems organized so that their constituent epistemic agents take on distinctive roles in a division of epistemic labor. This division of labor is as specified by systems of authority delegation. We now turn to a consideration of the role of tools and instruments in the scientonomic theory of the sociotechnical domain. The first discussion of the role of tools and instruments in scientonomic theory was authored by myself (Patton, 2019). Here, I will review the new sociotechnical elements introduced into scientonomic theory by my work with an attention to the role of tools and artifacts in distributed cognition.

Tool use occurs in a variety of animal species (Bentley-Condit & Smith, 2010). Following primatologist Jane van Lawick-Goodall, we can define tool use as the use of an external object as a functional extension of a body part in the attainment of an immediate goal (van Lawick-Goodall, 1970). In the cases we are concerned with here, the immediate goal in question is the acquisition, dissemination, or preservation of knowledge, and the tools we are concerned with are therefore called *epistemic tools* (Patton, 2019).

The simplest type of epistemic tool is a found object. Suppose that I encounter a pit. The bottom of the pit is hidden in darkness. I wish to answer the question 'How deep is that pit?'. Fortunately, I accept several theories that may be helpful in answering this question, which I represent diagrammatically in Figure 8.7.

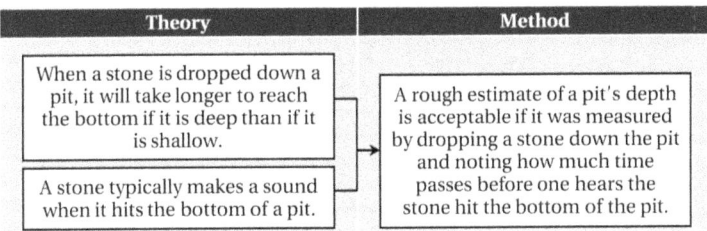

Figure 8.7: A theory-method diagram showing how an employed method is shaped by accepted theories

The theories assert a lawful relationship between the behavior of a stone and the depth of the pit. If they are correct and the lawful relationship does actually exist, then the behavior of the stone carries semantically meaningful information about the depth of the pit, and is a source of knowledge about it (Dretske, 1981, 1983). Using the theories, I formulate the method as stated in the diagram. In accordance with the law of method employment (Sebastian, 2016), the method is a deductive consequence of the theories. Under this method, dropping a stone down the pit, and noting the time interval between the release of the stone and the sound of its impact with the bottom is a normatively appropriate procedure for answering my question. The stone's role in a procedure that is a normatively appropriate way of answering a question given the agent's employed method is what makes the stone an epistemic tool (Patton, 2019).

According to Barseghyan (2015, pp. 7-8), employed methods of theory assessment form a hierarchy from abstract and general to concrete and specific. It is the more concrete levels of this hierarchy that specify epistemic tools and procedures for their proper use, by which acceptable answers to questions may be obtained. The resulting definition of *epistemic tool* is presented in Figure 8.8 (Patton, 2019).

Epistemic Tool ≡
A physical object or system is an epistemic tool for an epistemic agent *iff* there is a procedure by which the tool can provide an acceptable source of knowledge for answering some question under the employed method of that agent.

Figure 8.8: The definition of 'epistemic tool'

On scientonomic diagrams, epistemic tools are depicted by the symbol presented in Figure 8.9.

Figure 8.9: The diagrammatic symbol for depicting epistemic tools

9. The Diversity of Epistemic Tools and Their Role in Distributed Cognition

There are many kinds of epistemic tools. A found object, like the stone, is the simplest example, but epistemic tools are often manufactured artifacts designed to perform their epistemic function. Some of these artifacts simply extend our perceptual capacities. According to accepted acoustic theories, the stethoscope augments the capabilities of the human ear, allowing its user to hear faint sounds within the human body. There are procedures by which a skilled user can garner knowledge suited to answering a variety of questions about health and disease which will be acceptable under the currently employed methods of modern medicine. Windows, mirrors, microscopes, telescopes, and a blind person's cane also augment their users' senses.

Much as dropping as stone down a pit creates conditions that allow an epistemic agent to learn the depth of the pit, many other epistemic tools, such as an alchemist's furnace or the Large Hadron Collider, create special conditions or situations useful for scientific inquiry. The Large Hadron Collider, for example, is designed to accelerate beams of subatomic particles to speeds close to that of light and then cause the particles to collide with one another. According to the employed method of subatomic particle physics, data obtained from appropriate observations of such collisions are acceptable for testing the predictions of physical theories such as the standard model of particle physics (Mann, 2019).

It was once the case that spoken language was the only means by which propositions could be formulated and communicated. Human memory was once the only means by which they could be preserved. One subset of epistemic tools consists of those that represent propositions symbolically in a stable physical medium external to the body. This affords a number of means of cognitive enhancement, the simplest of which is the stable preservation of epistemic elements over time and their faithful transmission to other agents. The first such external representations were graphical (Donald, 1991, pp. 269-333). The oldest known human graphical markings where engraved on a rib 200,000 years ago. Graphical markings on bone or ivory did not become prevalent until 40,000 years ago. Various forms of writing and mathematical notation began to appear around 5000 years ago, and the phonetic alphabet, which allowed the graphical recording of spoken language, appeared about 3000 years ago. Clay tablets, written text on parchment or paper, hard drives, memory sticks, and various other data storage and transmission technologies allowed the creation and spread of symbols encoding the propositional knowledge of their authors.

This sort of epistemic tool, in which a physical medium external to the body is used to record symbols which convey epistemic elements, is a very important class of epistemic tool. However, such tools differ in important respects from our earlier examples, and this warrants further discussion. In Dretske's theory (Dretske, 1981, 1983), worldly events contain intentional semantic information about other worldly entities or processes because of their reliable causal relationships to them. An agent can extract this semantic content if their theories provide prior knowledge of these causal relationships. But while such causal relationships may account for the intentional content of the behavior of the stone about to the depth of the pit, or of subatomic particles in the Large Hadron Collider about the regularities posited by the standard model, written symbols don't typically stand in direct causal relationships to the entities or processes to which they refer. The semantic intentional content of such symbols is said to be conferred instead by the relationship in which they stand to an agent or agents who have knowledge of the meaning of the symbols (Jacob, 2019; Nanay, 2006). Given this significant difference, it's worth asking whether our current scientonomic definition of epistemic tool (Patton, 2019) is still applicable to them. By providing a scientonomic analysis of a simple representative example, I will show that it is. Consider a shopping list written with a pencil on paper. The diagram below shows that given accepted theories about shopping lists and written text, and employing a method deduced from them, a shopping list can provide an acceptable source of knowledge for answering a question, as required by our definition of epistemic tool (Figure 8.10).

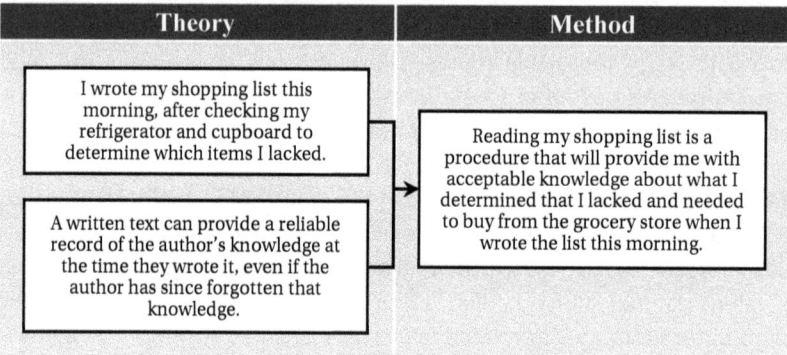

Figure 8.10: A theory-method diagram illustrating how a method can be employed that would allow a shopping list to provide an acceptable source of knowledge for answering a question

Additional features of our definition are probed by another example. In a pioneering work, Clark and Chalmers (1998) asked us to imagine a man, Otto, who carries a notebook with him to remember appointments because he is afflicted with Alzheimer's disease. A woman, Inga, uses neurons in her brain to accomplish the same purpose. They argue that since Otto's notebook performs the same function for him as Inga's neurons do for her that Otto's notebook should be considered a constituent part of his mind. Giere (2007) has advocated the more modest claim that Otto and his notebook form a distributed cognitive system. While accepting that Otto and his notebook form a distributed cognitive system, I wish to point out that the extended mind thesis conflates two different levels of cognitive organization. Because of this conflation, Otto's notebook satisfies our definition of an epistemic tool whereas Inga's neurons do not.

Inga's use of her neurons is not due to her own agency. It is due to the operation of learning processes that are sub-agential constituent parts of her and of which she is unaware. It is also due to evolutionary processes operating over a vast span of time prior to her existence. Because the functional role that Inga's neurons play in her ability to remember appointments is not due to her own agency, it cannot be said to be due to even the implicit acceptance by her of a theory or theories, or of any method deduced from them. This is why her neurons do not satisfy our definition of an epistemic tool. Otto, on the other hand, established his relationship to his notebook purposefully, as a whole behaving agent acting to solve a problem of which he was aware (perhaps with the guidance of other agents, such as his doctor, to which he delegated authority under theories he accepted). He could thus be said, at least implicitly, to semantically understand and accept a theory similar to the one which underwrites my use of the shopping list in the earlier example, from which a method follows under which the notebook is an acceptable source of knowledge. This is why Otto's notebook does satisfy our definition of an epistemic tool.

External representation in a stable physical medium is more than just a means of enhancing memory, it also is necessary in order for a human agent to perform any but the very simplest mathematical or logical computation. As we saw above, the sub-symbolic parallel distributed processing characteristic of the human brain is not well suited to the step-by-step manipulation of symbols according to rules (Rumelhart et al., 1986, pp. 44-48). Most people are lousy at doing math and logic in their head (Dehaene, 2011, pp. 104-128). Our capacities for doing so are vastly enhanced through the use of an epistemic tool like a pencil and notebook or a blackboard and chalk. These tools provide the needed external memory store for human epistemic agents to perform complex multistep symbolic computations (Clark, 2008; Clark & Chalmers, 1998; Dehaene, 2011; Giere & Moffatt, 2003; Rumelhart et al., 1986; Theiner, 2014). The great theoretical physicist Richard Feynman believed that his work took place, quite literally, on paper (Clark, 2008, pp. xxv-xxvi).

The contention that a human agent with a pencil and notebook, performing a mathematical computation, or making a shopping list, constitutes a distributed cognitive system is a highly reasonable one. The mathematical notation and physical diagrams in Feynman's notebook bore intentional semantic content for him. A variety of other physically comparable arrangements of graphite on paper would not. If Feynman's notebook were taken from him, he would have suffered a deficit in his mathematical cognitive abilities, just as surely as he would suffer some other sort of cognitive deficit if his brain were damaged. Feynman and his notebook formed a composed cognitive system in which each component performs a distinctive and complementary function in a division of cognitive labor. Feynman brought to the system a variety of capacities including pattern recognition, semantic understanding, creative insight, and agency. The notebook is capable of none of these things, but brings to the system a capacity that Feynman lacked, namely stable symbolic memory storage. Together the system had emergent cognitive abilities that either component, alone, lacked.

A pencil and paper provide an external memory store that allows a human agent to manipulate symbols according to rules. Computers perform the rule-based symbol manipulation itself. Computers clearly satisfy our definition of an epistemic tool, as there are a plethora of questions for which a computer, used in accordance with an appropriate procedure, can supply an acceptable source of knowledge under a method deduced from a theory appropriate to the question at issue. It would be possible, in principle, to describe the behavior of the computer purely in the language of solid-state physics. But such a description would leave one completely in the dark about why the machine is organized the way it is, rather than in some other way. This ontologically real state of organization can only be grasped by taking what Dennett (1987) called the design stance. A computer is carefully designed so that its inputs and outputs, and everything that happens in between, can be interpreted in a semantically meaningful way by human agents such as programmers or users, as the manipulation of symbols in accordance with rules. This special purposeful state of organization is what makes the aid that the computer provides cognitive rather than simply mechanical. The fact that the computer's rule-based symbol manipulation is quite different from the brain's sub-symbolic dynamical processing does nothing to weaken the claim that the computer and its user together constitute a distributed cognitive system. Such distinctive and complementary capacities are precisely what we might expect in a composed system with its division of cognitive labor.

Some symbolic epistemic tools, such as scales, thermometers, and voltmeters, are designed to produce mathematical symbols representing quantities as their output, thereby permitting quantitative measurement. A voltmeter's output, for example, may be read as the proposition that "this battery generates a voltage difference of 10 volts". Note the semantic content of such readings may be underwritten by Dretske's causal semantics (Dretske, 1981, 1983) since the position of the voltmeter's needle or the digits of its digital display are causally linked to the electrical potential difference between the positive and negative terminals of the battery. The semantic content of the symbols, as discussed above, requires another sort of explanation.

In many modern epistemic tools, quantitative measurement, digital computation, and a semantically interpretable output are combined. Consider a modern DNA sequencer (Heather & Chain, 2016; Hutchinson, 2007). When a prepared sample of DNA is inserted into such a machine the machine determines the order of the four nucleotide base pairs in the sample. This sequence carries the genetic code. The design and use of such a device draws on accepted theories from varied fields, from solid-state physics to molecular biology. The machine reports the base pair sequence as a text string on a computer screen. This output is full of intentional content for a molecular biologist with the appropriate knowledge. Note that the individual user of such a complex epistemic tool is almost certainly not familiar with all of the theories from which its acceptability as a means to acquiring reliable data is deduced. Its use thus typically involves authority delegation to communities familiar with those theories.

10. Epistemic Tool Reliance

Relationships of authority delegation link individual and communal epistemic agents together into larger communal epistemic agents, specifying their division of epistemic labor. What specifies the relationships between epistemic agents and epistemic tools? From what I have said above, it should already be evident that it would not be plausible to suppose that epistemic agents might delegate authority to epistemic tools. Our definition specifies that authority can only be delegated to an epistemic agent deemed to be an expert on some topic *x*. An epistemic agent, in turn, must, by definition, have a semantic understanding of the propositions that constitute the epistemic element in question and its available alternatives, and choose among them with reason. Such an agent, for example, should be capable of justifying its epistemic stances in terms of the relevant employed methods of the community, responding to all objections. It should be obvious, for example, that a voltmeter can't justify the claim that 'this battery generates a voltage of 10 volts'. That role could only belong to an expert familiar, to at least some degree, with the workings of the voltmeter.

The question of whether or not individual or communal epistemic agents ever delegate authority to epistemic tools can only be answered by empirical study of current and historical instances. The challenges typically confronting the users of an epistemic tool can be grasped by considering an incident from the history of radio astronomy (Burke-Spolaor et al., 2010; Gibney, 2016; Petroff et al., 2015). Human-made and natural interference, from varied sources, is a frequent problem for radio astronomers, requiring them to make interference sources a topic of inquiry. Given theories about such interference sources, astronomers can deduce methods of data assessment from them. Such methods ensure that only data derived from astronomical sources is accepted and data derived from interference is rejected. On numerous occasions, astronomers at the 64-meter diameter radio telescope dish at the Parkes Observatory in Australia detected a particular type of fast radio burst which they dubbed a *peryton*. Although these bursts resembled other bursts thought to be of astronomical origin, they also showed some features that, under accepted theories, were tell-tale signs of terrestrial interference. Five separate published research papers focused on identifying the possible source of the interference. A variety of possible theories, including lightning storms or other atmospheric phenomena were assessed and rejected. In the end, the theory that the radio bursts were due to a microwave oven in an observatory lounge became accepted. New methods and norms followed aimed at preventing astronomers from confusing microwave ovens with astronomical signal sources in the future (Patton, 2019).

In Patton (2019), I considered a number of possibilities and could find no instances that could be regarded as the delegation of authority to epistemic tools. As the *peryton* incident indicates, authority delegation seems instead to be to the expert users of the tool in question. Such experts are intimately familiar with theories of the use and pitfalls associated with the tool, and with the methods deduced from them for assessment of data derived from it.

Generally speaking, current and historical epistemic tools do not possess the cognitive wherewithal to be the objects of authority delegation as we have defined it here, since they cannot semantically understand a theory and its alternatives, and choose among them with reason. Cognitive scientists, neuroscientists, and their philosophical allies seek a naturalistic understanding of semantic understanding in terms of the structure and function of brains, and the relationship between organisms and their environment (Morgan & Piccinini, 2018; Neander, 2012). Semantic understanding of propositions is today considered a grand challenge by computer scientists (Embley, 2004). But while such technologies may be on the horizon, they have not yet, to any significant degree, been achieved.

Under our theory, epistemic tools do not possess the properties needed to be the objects of authority delegation. I have reviewed some empirical evidence that they are not, in fact, objects of authority delegation (Patton, 2019). I have therefore proposed a new relationship between epistemic tools and epistemic agents (Patton, 2019). This relationship is epistemic tool *reliance*. It is based on the relationship between epistemic tools and the concrete requirements of an employed method, as already explained. The definition of the term is presented in Figure 8.11.

Tool Reliance ≡
An epistemic agent is said to rely on an epistemic tool *iff* there is a procedure through which the tool can provide an acceptable source of knowledge for answering some question under the employed method of that agent.

Figure 8.11: The definition of 'tool reliance'

Note that, like authority delegation, epistemic tool reliance can be accounted for reductively in terms of acceptance of theories and employment of methods.

11. Conclusion

We have reviewed briefly the entities and relations currently posited by scientonomic theory for the sociotechnical domain. The sociotechnical domain consists of the set of all epistemic agents and epistemic tools. Epistemic agents are of two sorts, individual epistemic agents and communal epistemic agents. The definitions of each are summarized in the diagram. Two sorts of relationships are posited. The relationship of authority delegation exists between epistemic agents, and the relationship of tool reliance that exists between epistemic agents and epistemic tools. I have argued here that the elements of this proposed theory are amenable to empirical testing, refinement, or revision on the basis of historical and sociological research.

Bibliography

Azevedo, F. A., Carvalho, L. R., Grinberg, L. T., Farfel, J. M., Ferretti, R. E., Leite, R. E., & Herculano-Houzel, S. (2009). Equal Numbers of Neuronal and Nonneuronal Cells Make the Human Brain an Isometrically Scaled-up Primate Brain. *Journal of Comparative Neurology*, 513(5), 532-541.

Barseghyan, H. (2015). *The Laws of Scientific Change*. Springer.

Barseghyan, H. (2018). Redrafting the Ontology of Scientific Change. *Scientonomy*, 2, 13-38.

Barseghyan, H., Patton, P., & Shaw, J. (Eds.) (in press). *Visualizing Worldviews: A Diagrammatic Notation for Belief Systems*.

Bechtel, W. (1988). *Philosophy of Mind: An Overview for Cognitive Science*. Psychology Press.

Bechtel, W. (2008). *Mental Mechanisms: Philosophical Perspectives on Cognitive Neuroscience*. Psychology Press.

Bechtel, W. & Abrahamsen, A. (2002). *Connectionism and the Mind, Second Edition*. Blackwell Publishers Inc.

Bedau, M. A. (1997). Weak Emergence. In Tomberlin (Ed.) (1997), 375-399.

Beer, R. D. (2000). Dynamical Approaches to Cognitive Science. *Trends in Cognitive Science*, 4(3), 91-99.

Bentley-Condit, V. K. & Smith, E. O. (2010). Animal Tool Use: Current Definitions and an Updated Comprehensive Catalog. *Behaviour*, 147, 185-221.

Burke-Spolaor, S., Bailes, M., Ekers, R., Macquart, J.-P., & Crawford III, F. (2010). Radio Bursts with Extragalactic Spectral Characteristics Show Terrestrial Origins. *The Astrophysical Journal*, 727(1), 18.

Carruthers, P., Stich, S., & Siegal, M. (Eds.) (2002). *The Cognitive Basis of Science*. Cambridge University Press.

Cartwright, N. & Montuschi, E. (Eds.) (2014). *Philosophy of Social Science. A New Introduction*. Oxford University Press.

Cetina, K. K. (2009). *Epistemic Cultures: How the Sciences Make Knowledge*. Harvard University Press.

Churchland, P. S. (1989). *Neurophilosophy: Towards a Unified Science of the Mind/Brain*. MIT Press.

Churchland, P. S. & Sejnowski, T. J. (1992). *The Computational Brain.* MIT Press.

Clark, A. (2001). *Natural-Born Cyborgs?* Springer.

Clark, A. (2007). Curing Cognitive Hiccups: A Defense of the Extended Mind. *The Journal of Philosophy,* 104(4), 163-192.

Clark, A. (2008). *Supersizing the Mind: Embodiment, Action, and Cognitive Extension.* Oxford University Press.

Clark, A. (2010). Memento's Revenge: The Extended Mind, Extended. In Menary (Ed.) (2010), 43-66.

Clark, A. (2015). *Surfing Uncertainty: Prediction, Action, and the Embodied Mind:* Oxford University Press.

Clark, A. & Chalmers, D. J. (1998). The Extended Mind. *Analysis,* 65, 1-11.

Cohen, R. S. & Schnelle, T. (Eds.) (1986). *Cognition and Fact: Materials on Ludwik Fleck.* Springer.

Corradini, A. & O'Connor, T. (Eds.) (2010). *Emergence in Science and Philosophy.* Routledge.

Dayan, P. & Abbott, L. (2001). *Theoretical Neuroscience: Computational and Mathematical Modeling of Neural Systems.* MIT Press.

Dehaene, S. (2011). *The Number Sense: How the Mind Creates Mathematics.* Oxford University Press.

Dennett, D. C. (1971). Intentional systems. *The Journal of Philosophy,* 68(4), 87-106.

Dennett, D. C. (1984). *Elbow Room: The Varieties of Free Will Worth Wanting.* MIT Press.

Dennett, D. C. (1987). *The Intentional Stance.* MIT Press.

Dennett, D. C. (1991). Real Patterns. *The Journal of Philosophy,* 88(1), 27-51.

Dennett, D. C. (2003). *Freedom Evolves.* Penguin Books.

Donald, M. (1991). *Origins of the Modern Mind: Three Stages in the Evolution of Culture and Cognition.* Harvard University Press.

Donovan, S. K. (2016). The Triumph of the Dawsonian Method. *Proceedings of the Geologists' Association,* 127(1), 101-106.

Dretske, F. I. (1981). *Knowledge and the Flow of Information.* MIT Press.

Dretske, F. I. (1983). Precis of Knowledge and the Flow of Information. *The Behavioral and Brain Sciences,* 6, 55-90.

Embley, D. W. (2004). Toward Semantic Understanding: An Approach Based on Information Extraction Ontologies. *ADC'04: Proceedings of the 15th Australasian Database Conference,* 27, 3-12.

Fitch, W. T. (2008). Nano-Intentionality: A Defense of Intrinsic Intentionality. *Biology & Philosophy,* 23(2), 157-177.

Fleck, L. (1986). The Problem of Epistemology. In Cohen & Schnelle (Eds.) (1986), 79-112.

Fleck, L. (2012). *Genesis and Development of a Scientific Fact.* University of Chicago Press.

Fodor, J. A. (1975). *The Language of Thought.* Harvard University Press.

Fraser, P. & Sarwar, A. (2018). A Compatibility Law and the Classification of Theory Change. *Scientonomy,* 2, 67-82.

Friston, K. (2003). Learning and Inference in the Brain. *Neural Networks*, 16(9), 1325-1352.

Friston, K., FitzGerald, T., Rigoli, F., Schwartenbeck, P., & Pezzulo, G. (2017). Active Inference: A Process Theory. *Neural Computation*, 29(1), 1-49.

Fulda, F. C. (2016). *Natural Agency: An Ecological Approach*. Doctoral Dissertation, University of Toronto.

Fulda, F. C. (2017). Natural Agency: The Case of Bacterial Cognition. *Journal of the American Philosophical Association*, 3(1), 69-90.

Gibney, E. (2016). Mystery in the Heavens. *Nature*, 534(7609), 610-612.

Giere, R. N. (1992). The Cognitive Construction of Scientific Knowledge (Response to Pickering). *Social Studies of Science*, 22(1), 95-107.

Giere, R. N. (2002). Scientific Cognition as Distributed Cognition. In Carruthers, Stich, & Siegal (Eds.) (2002), 285-299.

Giere, R. N. (2002). Models as Parts of Distributed Cognitive Systems. In Magnani & Nersessian (Eds.) (2002), 227-241.

Giere, R. N. (2003). The Role of Computation in Scientific Cognition. *Journal of Experimental & Theoretical Artificial Intelligence*, 15(2), 195-202.

Giere, R. N. (2004). The Problem of Agency in Scientific Distributed Cognitive Systems. *Journal of Cognition and Culture*, 4(3), 759-774.

Giere, R. N. (2007). Distributed Cognition without Distributed Knowing. *Social Epistemology*, 21(3), 313-320.

Giere, R. N. (2010). *Explaining Science: A Cognitive Approach*. University of Chicago Press.

Giere, R. N. & Moffatt, B. (2003). Distributed Cognition: Where the Cognitive and the Social Merge. *Social Studies of Science*, 33(2), 301-310.

Godfrey-Smith, P. (2006). Mental Representation, Naturalism, and Teleosemantics. In Macdonald & Papineau (Eds.) (2006), 42-68.

Godfrey-Smith, P. & Sterelny, K. (2016). Biological Information. In Zalta, E. N. (Ed.) (2016). *The Stanford Encyclopedia of Philosophy (Summer 2016 Edition)*. Retrieved from: https://plato.stanford.edu/archives/sum2016/entries/information-biological.

Goldberg, S. C. (2016). A Proposed Research Program for Social Epistemology. In Reider (Ed.) (2016), 3-20.

Heather, J. M. & Chain, B. (2016). The Sequence of Sequencers: The History of Sequencing DNA. *Genomics*, 107(1), 1-8.

Herculano-Houzel, S. (2009). The Human Brain in Numbers: A Linearly Scaled-up Primate Brain. *Frontiers in Human Neuroscience*, 3, 31.

Hohwy, J. (2013). *The Predictive Mind*. Oxford University Press.

Hook, S. (Ed.) (1960). *Dimensions of Mind*. New York University Press.

Hutchinson, C. A. (2007). DNA Sequencing: Bench to Bedside and Beyond. *Nucleic Acids Research*, 35(18), 6227-6237.

Isaacson, W. (2007). *Einstein: His Life and Universe*. Simon and Schuster.

Jacob, P. (2019). Intentionality. In Zalta, E. N. (Ed.) (2019). *The Stanford Encyclopedia of Philosophy*. Retrieved from: https://plato.stanford.edu/archives/spr2019/entries/intentionality/.

Jacquette, D. (2006). Brentano's Concept of Intentionality. In Jacquette (Ed.) (2006), 98-130.

Jacquette, D. (Ed.) (2006). *The Cambridge Companion to Brentano.* Cambridge University Press.

Kelso, J. S. (1997). *Dynamic Patterns: The Self-Organization of Brain and Behavior.* MIT press.

Kim, J. (1999). Making Sense of Emergence. *Philosophical Studies,* 95(1), 3-36.

Latour, B. (1987). *Science in Action: How to Follow Scientists and Engineers Through Society.* Harvard University Press.

Latour, B. (2005). *Reassembling the Social: An Introduction to Actor-Network-Theory.* Oxford University Press.

Latour, B. & Woolgar, S. (1986). *Laboratory Life: The Construction of Scientific Facts.* Princeton University Press.

Lawson, T. (2014). A Conception of Social Ontology. In Pratten (Ed.) (2014), 19-52.

Lehrman, D. S., Hinde, R. A., & Shaw, E. (Eds.) (1970). *Advances in the Study of Behavior, Vol. 3.* Academic Press.

Levin, J. (2018). Functionalism. In Zalta, E. N. (Ed.) (2018). *The Stanford Encyclopedia of Philosophy (Fall 2018 Edition).* Retrieved from: https://plato.stanford.edu/archives/fall2018/entries/functionalism/.

List, C. & Pettit, P. (2006). Group Agency and Supervenience. *The Southern Journal of Philosophy,* 44(S1), 85-105.

List, C. & Pettit, P. (2011). *Group Agency: The Possibility, Design, and Status of Corporate Agents.* Oxford University Press.

Loiselle, M. (2017). Multiple Authority Delegation in Art Authentication. *Scientonomy,* 1, 41-53.

Longino, H. (1990). *Science as Social Knowledge: Values and Objectivity in Scientific Inquiry.* Princeton University Press.

Longino, H. (1996). Cognitive and Non-Cognitive Values in Science: Rethinking the Dichotomy. In Nelson (Ed.) (1996), 39-58.

Longino, H. (2019). The Social Dimensions of Scientific Knowledge. In Zalta, E. N. (Ed.) (2019). *The Stanford Encyclopedia of Philosophy (Summer 2019 Edition).* Retrieved from: https://plato.stanford.edu/archives/sum2019/entries/scientific-knowledge-social/.

Macdonald, G. & Papineau, D. (Eds.) (2006). *Teleosemantics: New Philosophical Essays.* Oxford University Press.

Machado-Marques, S. & Patton, P. (2021). Scientific Error and Error Handling. *Scientonomy,* 4, 21-39.

Magnani, L. & Nersessian, N. J. (Eds.) (2002). *Model-Based Reasoning: Science, Technology, Values.* Springer.

Mann, A. (2019). What is the Large Hadron Collider? *Live Science.* Retrieved from: https://www.livescience.com/64623-large-hadron-collider.html.

Mayr, E. (2006). Typological Versus Population Thinking. In Sober (Ed.) (2006), 325-328.

Menary, R. (Ed.) (2010). *The Extended Mind.* MIT Press.

Millikan, R. G. (1984). *Language, Thought, and Other Biological Categories: New Foundations for Realism.* MIT Press.

Morgan, A. & Piccinini, G. (2018). Towards a Cognitive Neuroscience of Intentionality. *Minds and Machines,* 28(1), 119-139.

Nanay, B. (2006). Symmetry Between the Intentionality of Minds and Machines? The Biological Plausibility of Dennett's Account. *Journal of Minds and Machines,* 16(1), 57-71.

National Institute of Standards and Technology (2018). The NIST Reference on Constants, Units, and Uncertainty. Retrieved from: https://physics.nist.gov/cgi-bin/cuu/Value?me|search_for=electron+mass.

Neander, K. (2012). Teleological Theories of Mental Content. In Zalta, E. N. (Ed.) (2012). *The Stanford Encyclopedia of Philosophy (Spring 2012 Edition).* Retrieved from: https://plato.stanford.edu/archives/spr2012/entries/content-teleological/.

Neander, K. (2017). *A Mark of the Mental: In Defense of Informational Teleosemantics.* MIT Press.

Nelson, J. (Ed.) (1996). *Feminism, Science, and the Philosophy of Science.* Springer.

Newell, A. & Simon, H. A. (1976). Computer Science as Empirical Inquiry: Symbols and Search. *Communications of the American Society for Computing Machinery,* 19(3), 113-126.

O'Connor, T. & Yu Wong, H. (2015). Emergent Properties. In Zalta, E. N. (Ed.) (2015). *The Stanford Encyclopedia of Philosophy (Summer 2015 Edition).* Retrieved from: https://plato.stanford.edu/archives/sum2015/entries/properties-emergent/.

Overgaard, N. (2017). A Taxonomy for the Social Agents of Scientific Change. *Scientonomy,* 1, 55-62.

Overgaard, N. (2019). *On the Collective Intentionality of Epistemic Communities.* Doctoral Dissertation, University of Toronto.

Overgaard, N. & Loiselle, M. (2016). Authority Delegation. *Scientonomy,* 1, 11-18.

Palermos, O. & Pritchard, D. (2016). The Distribution of Epistemic Agency. In Reider (Ed.) (2016), 109-126.

Palermos, S. O. (2011). Belief-Forming Processes, Extended. *Review of Philosophy and Psychology,* 2(4), 741-765.

Palermos, S. O. (2016). The Dynamics of Group Cognition. *Minds and Machines,* 26(4), 409-440.

Palider, K. (2019). Reasons in the Scientonomic Ontology. *Scientonomy,* 3, 15-31.

Patton, P. (2019). Epistemic Tools and Epistemic Agents in Scientonomy. *Scientonomy,* 3, 63-89.

Patton, P. & Al-Zayadi, C. (2021). Disciplines in the Scientonomic Ontology. *Scientonomy,* 4, 59-85.

Patton, P., Overgaard, N., & Barseghyan, H. (2017). Reformulating the Second Law. *Scientonomy,* 1, 29-39.

Petroff, E., Keane, E., Barr, E., Reynolds, J., Sarkissian, J., Edwards, P., & Burke-Spolaor, S. (2015). Identifying the Source of Perytons at the Parkes Radio Telescope. *Monthly Notices of the Royal Astronomical Society,* 451(4), 3933-3940.

Pratten, S. (Ed.) (2014). *Social Ontology and Modern Economics*. Routledge.

Putnam, H. (1960). Minds and Machines. In Hook (Ed.) (1960), 138-164.

Rabinovich, M. I., Varona, P., Selverston, A. I., & Abarbanel, H. D. (2006). Dynamical Principles in Neuroscience. *Reviews of Modern Physics*, 78(4), 1213.

Rawleigh, W. (2018). The Status of Questions in the Ontology of Scientific Change. *Scientonomy*, 2, 1-12.

Reider, P. J. (Ed.) (2016). *Social Epistemology and Epistemic Agency: Decentralizing Epistemic Agency*. Rowman & Littlefield.

Rumelhart, D. E. & McClelland, J. L. (Eds.) (1986). *Parallel Distributed Processing: Explorations in the Microstructure of Cognition Volume 2: Psychological and Biological Models*. MIT Press.

Rumelhart, D. E., Smolensky, P., McClelland, J. L., & Hinton, G. (1986). Schema and Sequential Thought Processes in PDP Models. In Rumelhart & McClelland (Eds.) (1986), 7-57. Cambridge, MA, London, UK: MIT Press.

Sarwar, A. & Fraser, P. T. (2018). Scientificity and the Law of Theory Demarcation. *Scientonomy*, 2(1), 55-66.

Schlosser, M. (2015). Agency. In Zalta, E. N. (Ed.) (2019). *The Stanford Encyclopedia of Philosophy (Winter 2019 Edition)*. Retrieved from: https://plato.stanford.edu/archives/win2019/entries/agency.

Searle, J. R. (1995). *The Construction of Social Reality*. Free Press.

Searle, J. R. (2006). Social ontology: Some basic principles. *Anthropological Theory*, 6(1), 12-29.

Sebastien, Z. (2016). The Status of Normative Propositions in the Theory of Scientific Change. *Scientonomy*, 1, 1-9.

Sober, E. (2006). Evolution, Population Thinking, and Essentialism. In Sober (Ed.) (2006), 329-359.

Sober, E. (Ed.) (2006). *Conceptual Issues in Evolutionary Biology*. Bradford Books.

Sprevak, M. & Kallestrup, J. (Eds.) (2014). *New Waves in Philosophy of Mind*. Palgrave Macmillan.

Theiner, G. (2011). *Res Cogitans Extensia: A Philosophical Defense of the Extended Mind Thesis*. Peter Lang.

Theiner, G. (2014). A Beginner's Guide to Group Minds. In Sprevak & Kallestrup (Eds.) (2014), 301-322.

Theiner, G., Allen, C., & Goldstone, R. L. (2010). Recognizing Group Cognition. *Cognitive Systems Research*, 11(4), 378-395.

Theiner, G. & O'Connor, T. (2010). The Emergence of Group Cognition. In Corradini & O'Connor (Eds.) (2010), 6-78.

Thompson, E. (2007). *Mind in Life: Biology, Phenomenology, and the Sciences of Mind*. Harvard University Press.

Tollefsen, D. P. (2004). Collective Epistemic Agency. *Southwest Philosophy Review*, 20(1), 1-12.

Tollefsen, D. P. (2006). From Extended Mind to Collective Mind. *Cognitive Systems Research*, 7(2), 140-150.

Tollefsen, D. P. (2014). Social Ontology. In Cartwright & Montuschi (Eds.) (2014), 85-101.

Tomberlin, J. E. (Ed.) (1997). *Philosophical Perspectives: Mind, Causation, and World*. Blackwell.

Tuomela, R. (2002). *The Philosophy of Social Practices: A Collective Acceptance View*. Cambridge University Press.

Turing, A. M. (1950). Computing Machinery and Intelligence. *Mind*, 59(236), 433-460.

UNESCO (2013). Facts and Figures: Human Resources. *The UNESCO Science Report, Towards 2030*. Retrieved from: https://en.unesco.org/node/252277.

van Lawick-Goodall, J. (1970). Tool Use in Primates and Other Vertebrates. In Lehrman, Hinde, & Shaw (Eds.) (1970), 195-249.

Varela, F. J., Thompson, E., & Rosch, E. (1997). *The Embodied Mind: Cognitive Science and Human Experience*. MIT Press.

Walsh, D. M. (2012). Mechanism and Purpose: A Case for Natural Teleology. *Studies in History and Philosophy of Science Part C: Studies in History and Philosophy of Biological and Biomedical Sciences*, 43(1), 173-181.

Walsh, D. M. (2015). *Organisms, Agency, and Evolution*. Cambridge University Press.

Walsh, D. M. (2016). Objectcy and Agency: Towards a Methodological Vitalism.

Walter, H. (2009). *Neurophilosophy of Free Will: From Libertarian Illusions to a Concept of Natural Autonomy*. MIT Press.

Wimsatt, W. C. (2006). Aggregate, Composed, and Evolved Systems: Reductionistic Heuristics as Means to More Holistic Theories. *Biology and Philosophy*, 21(5), 667-702.

Wimsatt, W. C. (2007). *Re-engineering Philosophy for Limited Beings: Piecewise Approximations to Reality*. Harvard University Press.

Chapter 9

General System-Theoretic Framework for Theories of Scientific Change

Ameer Sarwar

University of Oxford

Abstract: My aim in this chapter is to introduce the general system theory and to provide directions for research. One of the central issues in scientonomy is that its object of study is ill-defined. I will begin to approach this question by drawing on general system theory. In so doing, I will introduce the scientonomic community to a radically different way of thinking about explaining changes in scientific worldviews. Even if many of my ideas appear radical, I hope that by contradistinction the reader may appreciate how the scientonomic ideas may be made more precise.

Keywords: general system theory; explanation; operational closure; formal operationalization; scientific change

***

1. Introduction

All scientific fields begin by providing a general description of what they aim to study. This often involves specifying what the object of study – and the granularity of the level of analysis – is. This seems imperative because the adequacy of scientific explanations is difficult to assess without fully understanding the phenomena they are attempting to explain. This paper argues that the object of study of scientonomy has been inadequately specified. While it is no doubt difficult to precisely determine what "scientific change" amounts to, we surely need *some* understanding of what is being explained.

This chapter has three central aims. The first aim is to argue that the object of scientonomic study is insufficiently specific. The second is to get clearer on the various perspectives (static versus dynamic) that may be taken on the object of study, the different types of explanations (proximate versus evolutionary) that could be given, and the levels of analysis at which it may be studied. My final aim is to introduce general system theory as a radically different way of thinking about scientific change. Even if the technical apparatus of this approach is not incorporated in scientonomy, it will at least shed some light on how we may conceptualize complex change from a system-theoretic perspective. My hope is that the reader would appreciate the need for specifying the object of study, as well as the types of explanation and the levels of analysis, even if they think that my own approach ultimately fails.

2. Brief Historical Prelude

Following the so-called "historicist turn" induced by Kuhn's (2012) *Structure of Scientific Revolutions*, the role of history of science in philosophy of science became better recognized. It was further appreciated that describing the historical development of science is different from stipulating how it ought to be practiced. The former had previously been the purview of historical research, whereas the latter was in the domain of philosophical research.

Laudan recognized that the history and philosophy of science could give rise to a naturalistic discipline that explained the evolution of scientific worldviews overtime (Laudan, 1987, 1990; Donovan, Laudan, & Laudan (Eds.), 1988). Like other historical sciences, such as evolutionary biology or geology, this science would be evaluated in light of historical evidence. The history of science had become the object of scientific study.

The roots of *scientonomy* can be traced to Laudan's new science. Scientonomy aims to explain the process of scientific change (Barseghyan, 2015). Its laws are meant to be explanatory and naturalistic. They are explanatory in that they purport to uncover the general mechanism underlying scientific changes. They are naturalistic in that its theories are supposedly "tested" in light of detailed historical evidence. This ambitious new discipline aims to explain changes in scientific mosaics – the collection of theories, methods, and questions (Barseghyan, 2015, 2018).

3. The Three C's: Cause, Constitution, and Context

In this section, I will argue that scientonomy fails to specify its object of study, which is crucial for a naturalistic enterprise. There are at least two ways in which a science may fail in this regard. On the one hand, it may not provide a robust, formal definition of its object of study. This requirement, though ideal, is too restrictive for empirical science. Arguments about the precise definition of the object of study often devolve into semantic quibbles at the expense of genuine explanations.

On the other hand, a science may fail to specify what falls within its domain of inquiry. This does *not* mean that it fails to precisely define the object of study nor does it mean that it fails to explicitly list something that it tries to explain. Rather, the aim here is to have a clear understanding of what is being explained – this means that, minimally, the researcher can distinguish constitution (what something is), context (the environment in which it is situated), and cause (the explanans that explain the phenomena). It is in this latter sense that I think scientonomy does not sufficiently specify what is being explained.

A researcher may theoretically distinguish causally relevant factors from irrelevant ones only when she has a somewhat precise idea about what she is investigating. Thus, she may conceptualize a 'system' or 'object of study' as something distinct from, but nonetheless embedded in, a broader environment. The workings of the object under scrutiny may be studied reductively by decomposing it into its constituent parts and analyzing their interrelations. Alternatively, she may examine its relationship with the broader environment and explain how it is causally dependent on processes independent from it. These delineations are, of course, conceptual or methodological, not necessarily physical.

For instance, when studying the mechanisms of action potentials, a neuroscientist clearly identifies the cell membrane of the neuron as marking the system's boundary. Processes on one side of the membrane are clearly delineated as 'intracellular', while those on its other side are marked as 'extracellular'. The mechanism of action potential depends on ionic flow across the cell membrane. Without this boundary, it is not possible to explain the differential charges inside and outside of the cell prior to and after the action potential. Indeed, the causal relevance of ions in the extracellular space becomes clear only when we recognize that they are *not* part of the neuron. The neuron is, then, investigated as a 'system' with a determinate boundary and a surrounding environment that is causally relevant.

While it may be easier to delimit the cell given the topography of its membrane, a conceptual boundary is still necessary to study "scientific" change as opposed to, say, cultural or political change in the broader environment. As I mentioned, this need not be a precisely defined formal boundary. It could be a general description of what falls within the domain of inquiry, or somewhat more precisely, what exactly it is that we are trying to explain.

Regrettably, this is presently not the case in scientonomy. For example, in a recent paper introducing scientonomy, Rupik (2019) writes, "the *sociocultural factors theorem* states that these factors *can* affect theory assessment *provided they are part of the method of the time*" (p. 13, italics in original). Since sociocultural factors *can* affect theory assessment, there may also be times when they do not. We may ask scientonomists, "when do sociocultural factors affect theory assessment", and they may reply, "when the method of the relevant scientific community allows them to be". This does not seem like an adequate explanation, because one could simply change the concept of "the method of a scientific community" depending on whether sociocultural factors are explanatorily relevant. A more principled approach, in my view, would use sociocultural factors (if applicable) to explain scientific change without changing the definition of the scientific method of the time. This approach would, in other words, designate these factors as external processes that are causally relevant for explaining scientific change.

Let me provide a hypothetical example to motivate this point. It is uncontroversial that an instrument, such as a microscope or a telescope, is part of the scientific mosaic. It is based on accepted theories and generates reliable knowledge. It is typically used by experts in laboratories. However, its causal history can be traced further back. To begin, it had to be transported to the laboratory. It was manufactured using mined raw materials. Explaining how the microscope ended up in the biology laboratory requires us to invoke these factors. My question is: where do we draw the line? Is the microscope part of the scientific mosaic only when it is used in the lab? What about the expertise of the engineers who manufactured it? Are they relevant? The miners, too, have knowledge and skill that get utilized in producing the microscope. Are they, too, part of the scientific mosaic? Where, in this chain of causal events, is the boundary of the "scientific mosaic", which is the putative object of study? It appears to me that without a relatively clear understanding of the object of study, the scientonomic explanations, as it were, "spill over" to factors that are only distantly relevant.

We need to systematically distinguish cause, constitution, and context. Though I do not have a definition of *cause*, in this context all that matters is that certain aspects of the world feature into an explanation that accounts for the phenomena. A cause is something that, roughly speaking, plausibly explains aspects of its domain of inquiry without contradicting extant knowledge. What it means to plausibly explain something depends on the discipline and the expertise of its practitioners. 'Constitution' refers to the set of processes and components that form a system. They are literally the happenings (processes or functions) and the things (components or structures) that the system is composed of. The processes tend to be causes that are internal to the system. 'Context' refers to the environment in which the system is embedded. Some of the contextual factors are causally relevant while others are not. Because not all causal factors feature in an explanation, we first need to have a general understanding of the system under examination before we can systematically distinguish relevant factors from irrelevant ones.

Finally, one must be aware of the causal-constitution fallacy (Rupert, 2004). This fallacy holds that causality is sufficient for constitution – if *A* causally impacts *B*, then the former is thereby part of the latter. For example, even though the heart causally impacts the kidneys, the renal system is thereby not constitutive of the vascular system. The two systems are kept conceptually separate to examine their properties even though they are in reality intermingled. In light of this observation, one can separate causal factors that are part of the system from causal (and non-causal) environmental factors.[1]

4. Types of Explanations

I have thus far distinguished constitution (what something is), context (what it is embedded within), and cause (roughly, an explanation that provides reasons behind the phenomena). I now want to draw some further distinctions based on the work by Rahwan et al. (2019).[2] Consider a slightly modified version of their diagram presented in Table 9.1.

		Object of study	
		Dynamic	Static
Type of explanation	Proximate	Development (ontogeny)	Mechanism (causation)
	Evolutionary	Evolution (phylogeny)	Function (adaptive value)

Table 9.1: A modified version of a table presented in Rahwan et al., 2019

On the left side, we have the type of explanation or question. For most systems, such as biological systems or (as I maintain) scientific systems, one could ask at least two types of questions. The proximate question requires an explanation that appeals to causes that are "more recent". This often correlates with causes that are more recent in time, but I believe the appropriate interpretation is that proximate causes provide the minimal number of specific reasons that explain an object's current behavior. They do not, in other words, require one to invoke extremely general, historical reasons. An example of a proximate explanation for why, say, the biological community accepted the

---

[1] From a metaphysical perspective, no change in an object is detectable unless it is distinguished from its environment. From a methodological perspective, change in an object is difficult to detect without conceptually distinguishing it from its environment.
[2] Though the authors are working in the context of machine behavior, the conceptual apparatus they employ is explicitly in use in the biological sciences.

double-helix model of the DNA could be that they found the experimental work involving X-ray crystallography convincing.

By contrast, the evolutionary question necessarily invokes historical reasons. It asks, "how did the system come to assume its current form from the perspective of its (and its species') history?" In the example of the acceptance of the DNA, the question asks us to take a broader perspective in which we may need to explain the conditions that made its acceptance possible. One explanation could be that after Darwin's work on evolutionary theory, which provided a non-theological account of the origin of life, and the work of statistical geneticists in the early 20th century, which (to my understanding) took a frequency-centered approach to explaining inheritance across generations, the time was ripe for providing molecular underpinnings of these processes. An evolutionary explanation for the acceptance of the double-helix model, then, requires us to explain the conditions that made it possible for the scientific community of the time to accept it. I am inclined to think that evolutionary explanations are best appreciated by taking a counterfactual approach: the DNA model would likely not have been accepted at a time when the prevailing scientific opinion was that life is explained in terms of a non-physical "vital" force.

On the right side of the diagram, we have the object of study, which could be viewed either dynamically or statically. Of course, nothing is ever actually static. The idea behind taking a static perspective on the object of study is that we view it as a fully-formed, integrated whole at the end of its development. This is the hallmark of reductionist science, wherein, say, the biological organism is (often incorrectly) viewed as having reached its final stage of maturity. The proximate approach towards the static system asks us to explain the "recent" causes that explain its current state. A scientonomic example of this phenomena may be bibliometric analyses of citation trends that explain how a particular theory became accepted. Having said this, the mechanistic approach (providing a proximate explanation for a static entity) would often not apply to scientonomy given its focus on scientific *change*, which is inherently dynamical.

However, even within the static category, the evolutionary perspective may still be illuminating. This may be so because we may explain the development of science in terms of its functional value to society at large, e.g., by providing new technologies. This may explain the continued existence of science over centuries, though it is questionable whether it explains the change in scientific worldviews *per se*.

The most important column in this context is the dynamical view, according to which the researcher examines how the system changes. This is undoubtedly very complex in part because the scientist needs to determine the conditions that warrant referring to it as "the same system" over time. As before, we could provide proximate and evolutionary explanations when we view the system dynamically.

The proximate explanation concerns the recent development of the system by invoking "historically recent" causes. In the neural sciences, for instance, the developmental perspective concerns the development of the brain from prenatal stages to, say, early adulthood. Though the explanations here are still in many ways reductionistic in that they cite proximate biological causes (at the molecular, cellular, cognitive, behavioral, and even social levels), the object of study is the developing brain. In the scientonomic context, I am inclined to think that the unit of developmental study is a specific scientific theory, method, or question. This is because these may be more easily discernible, and therefore scientifically examinable. Explaining how a particular scientific theory or method became accepted – and then employed by the community – constitutes providing an explanation of its development.

By comparison, the evolutionary perspective on the dynamical system concerns the change in the *type* or species of the system overtime. I think here the evolutionary perspective must necessarily be at the level of the type of object under question – e.g., theories in physics over the past two centuries – rather than the specific object. This is because providing an explanation of how a discernible unit of analysis came to be constitutes a developmental explanation, not an evolutionary one. As I previously explained, the evolutionary explanations in this case are likely concerned with providing explanations of the conditions that brought something about or made it likely/possible. For example, in the case of the double helix model, the evolutionary explanation is Darwin's theory and the work of statistical geneticists. The developmental explanation is how, say, Crick and Watson used the work of Franklin and others to determine the structure of the DNA that convinced the biological community of the time.

One may object to my characterization by claiming that providing a developmental explanation necessarily excludes the evolutionary perspective and vice versa, because I have claimed that evolutionary explanations are not concerned with the specific entity (theory, for instance) that is under scrutiny. The objector here is correct insofar as the specific individual does not feature into an evolutionary explanation. However, this does not imply that the evolutionary perspective is thereby rendered irrelevant or that it no longer sheds any light on the specific object of study. On the contrary, the developmental approach must be construed within the backdrop of an evolutionary account. After all, the explanation behind the acceptance of the double helix model – or of why the synaptic connections undergo rapid proliferation followed by pruning – seems lacking without understanding the conditions that made the development possible in the first place. It is for this reason that I think the two approaches are complementary, not antithetical, to the study of changing systems.

In short, then, I believe that the phylogenetic (evolutionary) and the developmental (ontogeny) approaches are distinct but complementary in providing a fuller explanation of change, including scientific change. The former requires an ultimate explanation of how certain conditions came to be, whereas the latter gives proximate explanations for the development of a specific unit. Finally, the mechanistic and teleological (adaptive functionality) accounts may be relevant, though that requires taking a static perspective on systems whose changing nature we are interested in explaining.

5. Levels of Analysis and Approaches

In most sciences, the phenomena of interest can be viewed from multiple perspectives. In the physical or biological sciences, this often maps onto the physical sizes of the systems under question. In the neurosciences, for example, the primary object of study is the central nervous system. However, it can be analyzed at the molecular, cellular, circuit, systems, or even the organismal levels. All of these perspectives are mutually complementary: they add to our understanding of what the others may exclude. For instance, the cellular level explanations tell us how the glial cells (oligodendrocytes) support myelin sheaths of neurons within the brain. We could literally "zoom out" and consider the layers of the neocortex in which neurons of different varieties appear. Interestingly, the proportions vary depending on the neocortical regions from which the histological samples are taken. Specifically, if they are taken from brain regions that specialize in, say, processing sensory information, then the sensory neurons (granular cells) compose a greater proportion of the layer. But if the sample is from regions that specialize in movement, then the motor neurons (pyramidal neurons) constitute a greater proportion. My main reason for going into these details is to show that the same object of study can be viewed from different perspectives: we can take the cellular perspectives that explain the function and structure of individual neurons; we could similarly approach the issue from a histological perspective which tells us the relative compositions of neuronal layers; and, finally, we could look at entire brain regions and determine what they may be specialized in by considering the neocortical layers and the accompanying behavior of the organism.

In a like manner, I believe that "scientific worldviews" could be broken down into a number of levels, though they do not necessarily correspond with the physical sizes as we saw in neuroscience. Here is a tentative list: individual scientists, research groups/labs, departments, conferences, publications, scientometrics (citation trends), professional associations, funding agencies, governmental commissions, broader cultural or political trends, etc. These could be grouped along the semantic and non-semantic dimensions (Table 9.2).

Object of study: scientific change	
Semantic: scientific worldviews	Non-semantic: structural forces
Individual scientists	Departments
Groups/labs	Funding agencies
Conferences	Governmental commissions
Publications	Cultural or political trends
Professional associations	

Table 9.2: Semantic and non-semantic dimensions

Scientonomy's emergence from the history and philosophy of science context suggests that the original conception of "scientific worldviews" has to do with the semantic content – theories, methods, and questions – with which the scientists are concerned. From this perspective, the agents or units that are most likely "internal" to this system are individual scientists, research groups or labs, conferences, and publications. These are the types of things one associates working scientists with. These appear to be the appropriate types of units that need to be considered if we are interested in explaining the semantic content of scientific change.

However, as many have rightly pointed out, science does not operate in isolation from the broader society and scientists are subject to the same types of cultural forces that influence the general public. Nevertheless, I am not certain that simply pointing out, as Rupik does, that sociocultural factors can influence science "provided they are part of the method of the time" correctly analyzes the reasons as to why they could alter the semantic content of scientific worldview. It seems to me that these trends are external factors that nonetheless indirectly influence science. Instead of changing the semantic content of scientific theories per se, they change attitudes that may usher changes in the government, which may in turn lead to readjustments of the policies of the funding agencies in accordance with "areas of priority" or "national interest". Naturally, some types of research questions would receive less money than others, and so certain kinds of research would flourish while others will diminish. This process would ultimately bring about a change in the scientific worldviews but only indirectly: the cultural and political factors lead to a change in the scientific worldviews by influencing the non-semantic features that enable or inhibit certain semantic ones. Since here we are concerned with the semantic features as being 'internal' to science, we proximately explain the change in terms of the semantic content but evolutionarily explain it in terms of the sociocultural trends. In light of this discussion, it makes little sense to say that the sociocultural forces are "part of the method of the time".

I must emphasize that in this example, the evolutionary explanation maps onto the non-semantic factors, whereas the proximate one lies with the semantic ones. This is an accidental feature of the example, not a manifestation of an essential connection between these concepts. As I previously explained, the acceptance of the double helix model of the DNA can be evolutionarily understood in terms of the developments in Darwin's theory and statistical genetics, as well as proximately with respect to the evidence from X-ray crystallography. Both of these features are semantic. Accordingly, there is no connection between an evolutionary explanation necessarily mapping onto either the semantic or the non-semantic dimension. Similarly, the proximate explanation can map onto both the semantic and non-semantic aspects. Without belaboring the point any further, the evolutionary-proximate and the semantic-non-semantic categories are doubly dissociable.

Moreover, there are different research approaches one may take towards the same questions and levels of analyses. Table 9.3 summarizes some of the approaches that are conducive to this study, though there are likely many that I am excluding.

Research approach	Level of analysis	Type of explanation
Archival research	All	Largely evolutionary
Interviews	Individuals or groups	Proximate
Quantitative Semantic analysis	Published material	Both
Scientometrics	Publications, funding data, government decisions, etc.	Largely Proximate

Table 9.3: Some of the approaches towards the same questions and levels of analyses

I will briefly explain scientometrics to demonstrate how it maps onto the semantic and non-semantic aspects of scientific change, as well as illustrate its relationships with the relevant level of analysis and type of explanation. Scientometrics is a subfield of bibliometrics concerned with statistically analyzing trends in scientific publications, especially with reference to citation patterns. This perspective illuminates the structure of science: it tells us what kinds of questions the scientists are studying, which kinds of methods are being employed, and who is considered an authority on specific topics. This approach is non-semantic in the sense that here researchers are not discussing the content of specific scientific theories. They are not, say, explaining whether a paper was retracted because it utilized a highly faulty statistical technique. Rather, the focus is on analyzing patterns without reference to the content of each data point with the hope of finding meaningful trends.

Why, you may ask, are these researchers finding meaningful trends? This is because they are interested in studying how the semantic content changes over time. By analyzing the trends in, say, the types of questions that are being pursued, it may be possible to predict that the future answer that becomes accepted would noticeably change the "worldview" of the specific field. In this way, the scientometrics approach, while not explicitly referring to the semantic content of the research articles, can predict and explain how the worldview may change.

The scientometrics approach almost by definition works with large datasets, so it inevitably involves data on funding, governmental decisions, publications, citations, etc. A common feature between these things is that they are all recent. Because of their recency, the kinds of explanations scientometrics offers are largely proximate: it explains how specific theories may become accepted without alluding to centuries-old facts that changed the scientific worldview. This is not because of an inherent conceptual limitation of the method. The problem is rather the unavailability of large datasets from the distant past. However, if archived works could be converted into large datasets using semantic analyses (using machine learning), then it may be possible to conduct scientometric analyses on them.

In this light, I want to strongly emphasize that methodological pluralism is probably the best approach when studying a topic as complex as scientific change. A researcher should simply use a method that best answers specific questions while being cognizant of its limitations. The usage of various methods on related questions may illuminate aspects that the other methods exclude, as well as allow us to provide different types of explanations that give a fuller picture of the target phenomena.

In this section, I have tried to divide the object of study along the semantic and non-semantic dimensions. The former has to do with the content of scientific theories, methods, and questions, whereas the latter concerns with larger forces that may indirectly influence scientific change. It also includes unique approaches, such as scientometrics, that researchers may take to study specific phenomena. It is crucial that one reflects on the limitations of the methods and the kinds of explanations they allow.

activities would likely continue. Therefore, if enabling conditions cease to exist, then the processes they enable also stop; by comparison, if effectuators do not exist, then the processes they impact may slow down but do not stop. As these examples show, enabling conditions and effectuators are graded rather than binary.

8. Operationalizable Variables

I will now consider seven variables that may be operationalized based on the relationships explained above. I will briefly outline the implications of each.

First, the amplitude or strength of connection is the causal impact of one process on another. In high-amplitude relationships, altering the antecedent process greatly impacts – either positively or negatively – the subsequent process. High-amplitude relationships potentially have ripple effects depending on the amplitude and number of effectuator relationships supported by the subsequent process (and so forth). Second, the frequency refers to the number of times an effectuator connection is realized over a specified period of time. Third, duration is the length of time over which a given connection is active. Thus, while frequency corresponds with the number of times a connection is active, duration corresponds with the length of time for which each of these is active. Enabling conditions have meaningful frequency and duration measures only when we measure the *change* in their values. For instance, we may wish to know how frequently or for how long an enabling condition is above a certain value. This is required because enabling conditions cannot have their value at zero. By comparison, effectuator relationships can have frequency or duration measured in light of their values being zero or non-zero.

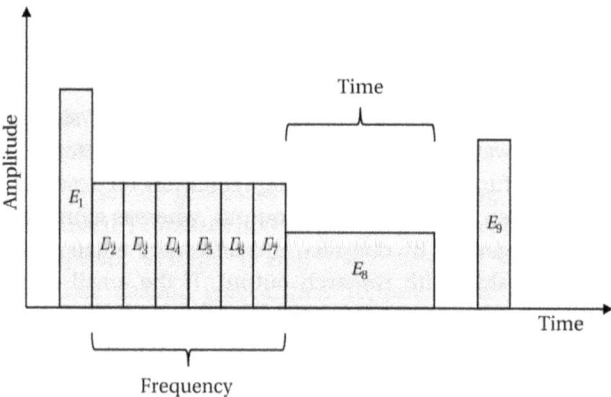

Figure 9.2: A graph showing the relations between strength, frequency, and duration

Before explaining the remaining variables, I wish to explain how strength, frequency, and duration are related. Let us consider a two-dimensional graph with a measure of "strength" on the y-axis and unit of time on the x-axis. In the example of research output, the y-axis may contain the number of books or articles published, or the number of citations generated. The x-axis may have years or decades. We may produce a graph, like the one below, in which the height of each bar corresponds with its amplitude, the number of bars overtime denotes frequency, and the length of each bar suggests duration (Figure 9.2).

Of course, these bars would likely overlap, but here they are represented as being mutually exclusive for the purposes of representation. This sketch is a very basic starting point and is far from complete. The important point is that we can make some headway into quantitatively studying scientonomic phenomena once we precisely define our concepts and object of study. Four features of the figure need to be clarified. First, the gap between E_8 and E_9 represents the absence of effectuators. This likely will not occur in practice, but it needs to be shown in principle. Second, the frequency is most meaningful when the effectuators (here, from E_2 to E_7) are virtually identical to each other. This does not, however, imply that frequency information involving non-identical effectuators is vacuous. Third, two methods can be used to calculate means (average): (a) diving the aggregate amplitude by the number of effectuators; (b) dividing the aggregate amplitude by the total length of the effectuators. Finally, the area covered by an effectuator (amplitude x length) may quantify its overall "impact".

The three variables I have discussed pertain to the relationship between any two processes. The next four variables concern the entire system.

Fourth, the number of relationships in a system is the aggregate of all the connections between all of the processes. An analogous concept is used in economics, wherein the aggregate demand and supply curves at the macroeconomic level are the aggregated results of individual demand and supply decisions, respectively.

Fifth, the processing in the system may occur serially or in parallel. The activity of enabling conditions is always in parallel to ensure operational closure. But the activity of effectuators can be parallel or serial. In a scientific mosaic, for example, the activity is in parallel: the scientists are producing papers, the journals are publishing them, and the students are studying them.

Sixth, distribution is the spread of strength, frequency, and the number of relationships between processes. Various topographical arrangements would indicate different types of systems. For example, a system in which most of the processes emanate from a few central ones may be characterized as unipolar or hierarchical. A system with two "poles" of heightened activity is bipolar.

Interestingly, if the activity of the system is mapped over time, we may find that systems exhibit bipolarity before a mosaic split. In other systems, processes may be distributed more uniformly, thereby indicating decentralized organization. Once the distributions of strength, frequency, and number are mapped, they may be superimposed. If they overlap to uncover the "hot spots" of activity, we would have convergent evidence of what a given community under investigation deems as important.

Finally, interdependence is a consequence of distribution. The more uniformly distributed the connections, the more interdependent the system is. The first implication of higher uniformity is that ripple effects in one part can be absorbed by other parts. By comparison, this feature is not present in unipolar or bipolar systems. Rather, their ability to accommodate perturbations crucially depends on their locus of origin. If the problems emerge in the peripheral, non-polar regions of the system, then they can be easily accommodated. However, if these emerge in one of the central processes, then the peripheral processes likely cannot accommodate them. The second implication of distribution is that a system's processing efficiency may depend on its polarity, though I am not certain whether uniform systems are more efficient than polar ones. Nonetheless, in all kinds of systems, redundancy increases as systems grow larger. Redundancy corresponds with the proportion of processes or effectuators that have been duplicated. These processes may very well be inefficient. However, they may be "stand ins" for the current processes in case they malfunction. This adds a layer of security to the system. Consequently, the cost of safety is inefficiency.

I will conclude this section with a cautionary remark: I have mostly outlined the positive aspects of each of the variables. Any number of combinations are possible that could lead to drastically negative effects in actual systems. Nonetheless, the most important point is that these variables – and, likely, very many others – may be operationalized to start providing more precise explanations of evolving systems, including systems of scientific knowledge.

9. Formal Properties and Inter-Systemic Relationships

This section extends the discussion by examining the formal relation of transitivity exhibited by systems, processes, and environment. I will also provide conditions for subordinate systems.

Operationally closed enabling conditions exhibit transitivity: process P_1 enables process P_2, which enables process P_3, and so P_1 also enables P_3. Because a system is a set of operationally closed enabling conditions, it is transitive. Moreover, transitivity also exists between processes that belong to the system and its environment. For example, process P_4 may enable process

E_1, which may enable process E_2. Here, P_4 also enables E_2. The environmental processes may likewise enable the system's processes. Importantly, the organization of these processes would not constitute organizational closure. Of course, not all enabling conditions are identical. Their unifying feature is that they are necessary conditions for the processes to which they give rise. Therefore, strictly speaking, the transitivity obtains between enabling conditions *qua* enabling conditions.

However, in the case of effectuators, only "weak transitivity" or "relation of similarity" is observed (Borgatti & Halgin, 2011): processes A and B are related by an effectuator connection, and B and C are related by another connection, so A and C are also related. The reason this relationship is non-transitive is that the A-B and B-C connections need not be identical. Indeed, it is unlikely that two effectuator connections are identical.

I will now briefly discuss the conditions for subsystems and supersystems. Let us have a system S_1 with at least three processes P_1, P_2, and P_3. Two of these processes, P_1 and P_2, are mutually enabling. These processes therefore satisfy the conditions of operational closure. Hence, they constitute a system S_2. There are at least two ways of construing the relationship between S_1 and S_2. First, S_1 is the environment within which S_2 is embedded. This implies that there must not be operational closure between S_1 and S_2. But this violates our stipulation that P_1 and P_2 are constitutive processes of S_1. Hence, we cannot think of S_1 being S_2's environment, though we may do so only metaphorically.

The second interpretation is that S_2 provides enabling conditions for S_1 while being operationally closed. Accordingly, S_2 is an independent system embedded within a larger system. Specifically, S_2 is a subsystem of S_1. Formally, a set of operationally closed enabling conditions, S_2, is a subsystem of S_1 if and only if these processes enable or are enabled by a subset of S_1's processes. In other words, S_2 is a subsystem of S_1 just in case all of its processes provide enabling conditions for or are enabled by a subset of S_1's processes. Examples may include organelles in cells, organs in the human body, physics departments in a university – all of these subsystems are operationally closed and provide enabling conditions for their superordinate systems.

A corollary of this formalization is that a set of processes that provide enabling conditions to a system cannot itself be called a subsystem if it is not operationally closed. This provides the researchers with formal criteria for distinguishing the causal impact of some process(es) on another set of processes, which may be organized in a system, without thereby thinking that the former process(es) become part of the system to which the latter processes belong. Specifically, construing complex phenomena at the level of general system theory prevents one from making the causal-constitution fallacy.

To conclude, my purpose in this chapter has been to present, in outline, a general conception that may motivate the need to distinguish the intended object of study from background noise. Of course, there are all kinds of questions about "ecological validity" or whether systems are natural kinds or theoretical constructs. My hope has been to give the reader some material with which to think about the issues at hand: what are the principles based on which we isolate the object of study from everything that causally impacts it but is not a part of it? This question, more so than my attempted answer, which may be developed into a theory by those sufficiently inclined and far more talented, is what I hope leaves the reader puzzled after finishing this chapter.

Bibliography

Barseghyan, H. (2015). *The Laws of Scientific Change*. Springer.

Barseghyan, H. (2018). Redrafting the Ontology of Scientific Change. *Scientonomy*, 2, 13-38.

Borgatti, S. P. & Halgin, D. S. (2011). On Network Theory. *Organization Science*, 22(5), 1168-1181.

De Jaegher, H., Di Paolo, E., & Gallagher, S. (2010). Can Social Interaction Constitute Social Cognition? *Trends in Cognitive Sciences*, 14(10), 441-447.

Di Paolo, E. & Thompson, E. (2014). The Enactive Approach. In Shapiro (Ed.) (2014), pp, 68-78.

Donovan, A., Laudan, L., & Laudan, R. (Eds.) (1992). *Scrutinizing Science. Empirical Studies of Scientific Change*. The Johns Hopkins University Press.

Dupré, J. (1983). The Disunity of Science. *Mind*, 92(367), 321-346.

Herring, E., Jones, K. M., Kiprijanov, K. S., & Sellers, L. M. (Eds.) (2019). *The Past, Present, and Future of Integrated History and Philosophy of Science*. Routledge.

Kuhn, T. S. (2012). *The Structure of Scientific Revolutions: 50th Anniversary Edition*. University of Chicago Press.

Laudan, L. (1987). Progress or Rationality? The Prospects for Normative Naturalism. *American Philosophical Quarterly*, 24(1), 19-31.

Laudan, L. (1990). Normative Naturalism. *Philosophy of Science*, 57(1), 44-59.

Port, R. F. & Van Gelder, T. (1995). *Mind as Motion: Explorations in the Dynamics of Cognition*. MIT Press.

Rahwan, I., Cebrian, M., Obradovich, N., Bongard, J., Bonnefon, J.-F., Breazeal, C., Crandall, J. W., Christakis, N. A., Couzin, I. D., Jackson, M. O., Jennings, N. R., Kamar, E., Kloumann, I. M., Larochelle, H., Lazer, D., McElreath, R., Mislove, A., Parkes, D. C., Pentland, A., Roberts, M. E., Shariff, A., Tenenbaum, J. B., & Wellman, M. (2019). Machine Behaviour. *Nature*, 568(7753), 477-486.

Rupert, R. D. (2004). Challenges to the Hypothesis of Extended Cognition. *The Journal of Philosophy*, 101(8), 389-428.

Rupik, G. (2019). Scientonomy: A Bold New Vision for an Integrated History and Philosophy of Science. In Herring et al. (Eds.) (2019), 19-37.

Shapiro, L. (Ed.) (2014). *The Routledge Handbook of Embodied Cognition*. Routledge.

Von Bertalanffy, L. (1950). An Outline of General System Theory. *British Journal for the Philosophy of Science*, 1, 134-165.

Von Bertalanffy, L. (1973). *General System Theory: Foundations, Development, Applications*. Penguin University Books.

Chapter 10

Pluralism in Scientonomy's Mechanism of Compatibility: Incompatible Pluralism of Mosaics and Pluralism of Compatible Theories

Deivide Garcia da S. Oliveira

Federal University of Recôncavo of Bahia

Abstract: This paper explores the relationship between pluralism and scientonomy, i.e., the ways in which scientonomy is or is not pluralist. Scientonomy claims by its zeroth law (the compatibility law) that the elements of the mosaic must be mutually compatible at any time. However, when Harder (2013) proposed to change the law of compatibility, he did it by answering the question of whether two or more elements can coexist in the same mosaic. He said that they can. However, this answer requires a detailed analysis of compatibility's meaning and its relation to the nature of the mosaic. That detailed analysis leads us to assess scientonomy by contrasting its pluralist and monist aspects. We will argue that scientonomy embraces some aspects of monism, as well as some aspects of pluralism. As a result, we argue that an adequate form of pluralism is the so-called *compatible pluralism of theories*, seen from a perspective internal to a scientific mosaic. A second form of pluralism becomes evident from a perspective outside any particular mosaic. It is what we will call the *pluralism of incompatible mosaics*. To explore these topics, after a brief description of the problem, we will review the main features of the zeroth law. In section three, we will briefly review some forms of pluralism to argue why one of them fits scientonomy better. In the end, in section four, will discuss the variety of monist and pluralist features of scientonomy, focusing on pluralism.

Keywords: pluralism; monism; compatibility; mosaic; methodology

***

1. Introduction

Scientonomy is a field of inquiry addressing questions about scientific change. It asserts that scientific theories change by a law-governed process. It has so far proposed four general laws. The first one is the *law of scientific inertia*, the second is the *law of theory acceptance*, the third is the *law of method employment*, and the zeroth law is the *law of compatibility*. This last law claims that theories in any given mosaic are mutually compatible. By doing so, it invites further inquiry about the conditions for compatibility. We accept this invitation through an investigation of some pluralist and monist aspects of the zeroth law and the mechanism of compatibility in the mosaic. That is, about the nature of compatibility as a whole in scientonomy's view on theory change.

The conditions under which elements in the mosaic are considered compatible can be explored in many ways. Amongst those ways, we chose the debate between pluralism and monism. This debate has a long history dealing with the consistency or inconsistency or the compatibility or incompatibility of theories dealing with the same domain or phenomenon. We will determine if, and in what sense, scientonomy's compatibilism is related to pluralism.

The remainder of this text will consist of three sections. In section 2, I will describe the law of compatibility and how the mechanism of compatibility deals with the coexistence of multiple theories in the mosaic. That description opens doors for our inquiry into pluralism and how scientonomy tackles it. I will explore some forms of pluralism in section 3. Once there is a plurality of pluralisms available on our supermarket shelves, we are going to pick one. In section 4, the problem of plurality, as raised and developed in sections 2 and 3, is explored and taken to its ultimate consequences. At the end of section 4, I present and discuss some limitations of compatibility and the zeroth law. Namely, I will inquire how scientonomy deals with complex cases where more than one accepted and incompatible theory answers questions in the same domain (*mosaic split*). Section 5 will offer some final considerations.

2. On the Mechanism of Compatibility in the Zeroth Law

Scientonomy's view of scientific change proposes that all changes of scientific theories and methods are by means of an underlying unchanging mechanism. Scientific change is a law-governed process where the laws explain the process of change. Scientific methods are a deductive consequence of theories and methods accepted and employed at the time, under the third law of scientific change (Sebastien, 2016).

The zeroth law states that "at any moment of time, the elements of the scientific mosaic are compatible with each other" (Barseghyan, 2015, p. 153; Harder, 2013). What kind of compatibility is the zeroth law referring to? How could this be approached from a pluralistic viewpoint? By answering these questions, we hope to open new doors to other research about other topics critical to the scientonomy community, such as the notion of the scientific mosaic,[1] and under what conditions there is compatibility among its elements.

However, before anything else, let us say something about the reason for calling it the 'zeroth' law instead of the 'fourth'. This is itself a philosophical matter. According to Barseghyan (2015), the reason is due to the *synchronic nature* of the zeroth law, for it "applies to the mosaic viewed from a *static* perspective". That is, it applies with respect to the compatibility of the elements within the mosaic if we could 'pause' the process of scientific change (Barseghyan, 2015, p. 153). So what Barseghyan is discussing is that when we look at the process of scientific change, and we pause that process, what we realize is that the mosaic's criteria for acceptability of theories and methods, for abstract and empirical reasons, does comprise compatibility, at least, in principle (Barseghyan, 2015, p. 153).

Although the zeroth law fosters compatibility, there is space to investigate under what specific conditions two or more inconsistent theories and methods, in the empirical sciences, can coexist inside a scientific mosaic. In contrast, in some other cases, contradictory theories and methods are not only inconsistent but incompatible, bringing up a problem for compatibility within a single mosaic.

Another thing to observe is that the criteria of compatibility change between times and fields of inquiry. But the role of compatibility criteria is, in principle, unchangeable – they are employed by an epistemic agent to determine whether a pair or epistemic elements can coexist within their mosaic. In contrast, the specific criteria of compatibility are altered in accordance with the third law as a result of changes in accepted theories at the time and field of inquiry. According to the second law, for any new theory to be accepted, it must satisfy the criteria already in place in the mosaic at the time. Furthermore, according to the first law, once an element is accepted, it remains in the mosaic until another element replaces it (Barseghyan, 2015, p. 124). Of course, the laws of scientific change do not exclude the possibility that two or more theories can be, simultaneously, accepted as the best available answers to their particular questions about the same phenomenon. That is, a single mosaic can hold one

---

[1] The scientific mosaic is defined as "a set of all epistemic elements accepted and/or employed by the epistemic agent" (Barseghyan, 2018, p. 36).

or more theories explaining different features of the same phenomena in a mutually compatible way.

On this account, the zeroth law, *in principle*, knowingly allows potential classical logical inconsistencies between two accepted theories that address the same phenomenon (Barseghyan, 2015, pp. 153-156). A famous example of this, mentioned by scientonomists, is the case of quantum theory (QT) and the theory of general relativity (GR) when both are applied to singularities in black holes. In this case, GR and QT are accepted as "the best available description *of their respective domains*" (Barseghyan, 2015, p. 154, italics added) even though their accounts of the same phenomenon (a black hole) are inconsistent.

A corollary of this is that contradictions and problems of compatibility, in practice, only come to light when we seek the *best* possible explanation of the same phenomenon, so we end up needing to be tolerant, even if only temporarily. The inconsistencies between two theories only occur in empirical sciences and in scenarios which, due to the conjunction of inconsistencies, we also embrace the complexity of some phenomena and necessity of multiple approaches. Thus, the contradiction between those theories, says Barseghyan (2015, p. 154), becomes apparent only when we apply them in conjunction with the same phenomenon seeking the best available explanation. According to scientonomy's view of scientific change (Barseghyan, 2015, p. 158), scientists may not like inconsistencies, but from an empirical viewpoint, sometimes there is no other option.

That being said, revealed inconsistencies do not lead to an endorsement of that contradiction. Rather, when scientists accept two theories of the same object, and the theories' contradiction becomes apparent, scientists try to "eliminate" it (Barseghyan, 2015, p. 158). Regardless of this drive, theories can simultaneously be compatible and inconsistent, because the "mutual compatibility of two theories is not necessarily decided on the basis of their logical consistency" (Barseghyan, 2015, p. 158). This distinction between compatibility and logical inconsistency is the key idea of the zeroth law.

Despite scientists' dissatisfaction with contradiction, the nature of our theories and methods of the contemporary scientific mosaic rejects infallibilism and embraces fallibilism, paraconsistent logic, and the view that theories should be seen as quasi-true (Barseghyan, 2015, pp. 154-155). Morrison, in her paper about scientific model inconsistency and complementarity, says that our search for absolutely 'complete' models should be replaced for a search "of contingency and 'good fit' to some specific aspect of a system" (Morrison, 2011, p. 343).

In a word, the adoption of fallibilism is in harmony with our current situation in science. Thus, to tackle scientific mosaics from the standpoint of compatibility, we must shift our view on the nature of theories from them being absolutely true to truth-like, and shift our attitude from inconsistency intolerance to inconsistency tolerance. Once we make these shifts, the viability of *fallible compatibilism* is achieved, at least in principle (Barseghyan, 2015, p. 156).

However, we could ask now about the limits of such compatibility, and such tolerance for coexistent contradictory theories. The short answer is that the mechanism of compatibility seems to be restricted. As we will explore, the cost of taking up compatibilism for Barseghyan's TSC involves the adoption of constraints guiding what aspects of compatibilism and inconsistency are tolerable, and which are not.

There are three aspects of compatibility. The first aspect is the compatibility of accepted theories in the *same mosaic*. This aspect takes place in two situations which I will call 1a and 1b, both derived from a general principle stating "two *theories* simultaneously accepted in the *same mosaic* cannot be incompatible with one another" (Barseghyan, 2015, p. 157, emphasis added). Thereby, in the first situation (1a), usually when two mutually inconsistent propositions are accepted into the mosaic, they "do not have the same object. More specifically, two propositions seem to be considered compatible by the contemporary community when, by and large, they explain different phenomena" (Barseghyan, 2015, p. 158). In this inconsistency-tolerant situation (1a), what happens is that, while looking for the best ways to eliminate inconsistencies, the scientific community can ask "whether the theories can be limited to their specific domains" (Barseghyan, 2015, pp. 158-159).

In a word, the mechanism is, in reality, talking about a specific type of conflict between two theories. That is why "the contemporary community is prepared to tolerate formal logical inconsistencies in the mosaic" (Barseghyan, 2015, p. 158). As an example of this scenario, Barseghyan (2015, p. 159) refers to the case of QT and GR. It does not matter if those theories are strictly inconsistent, from a classical logical view, for what matters is that they conflict only about such phenomena as singularities in black hole. Since QT and GR are accepted due to their success in explaining their respective domains of phenomena – the quantum/micro scale and macro scale respectively – scientists accept both theories, whereas *currently* and *temporarily* limiting the domains of applicability of the two theories with respect to singularities in black holes.

The second situation (1b) of the first general scenario of the mechanism of compatibility, is the so-called anomaly-tolerant situation. In this situation, Barseghyan says that the community can tolerate anomalies but between an "accepted *general* theory and a *singular* proposition describing some anomaly" concerning the same phenomenon where "the latter describes a counterexample for the former" (Barseghyan, 2015, p. 160, emphasis added). That is interesting because it indicates restrictions about tolerance, and those restrictions will help us find what kind of pluralism fits better for scientonomy.

The second general aspect of the compatibility law has to do with methods, and states that "two simultaneously employed methods cannot be mutually incompatible" (Barseghyan, 2015, p. 162). So, unless these methods state necessary and sufficient conditions for theory acceptance, two methods can be mutually compatible, and even complementary. That can be true also when they have different (but not conflicting) assessments about theory acceptance (A *AND* B), and when their employment requires a disjunctive application (option A *OR* B) (Barseghyan, 2015, p. 163). This aspect tolerates more than one or two methods in the same mosaic under the condition that they are not exhaustive or when their employment does not happen simultaneously.

The third and last aspect of the law of compatibility states "that a method and a theory that are mutually incompatible cannot simultaneously be in the same mosaic", i.e., a method (prescriptive) cannot conflict directly with a theory (descriptive) (Barseghyan, 2015, pp. 163-164). The direction of assessment flows from theory to methods since "*the third law* stipulates that our accepted theories shape our employed methods" (Barseghyan, 2015, p. 132, emphasis added). Nonetheless, neither the second or third aspects will be addressed explicitly in section 4 when we synthesize sections 2 and 3, as there is no need to do it for discussing the question of pluralism in scientonomy.

Our discussion of the zeroth law's general aspects raises some questions that we will explore in sections 3 and 4. The first issue is about some general form of pluralism versus monism (oneness) in the sciences. To do that, we will consider some specific schools of pluralism, including the Stanford and Minnesota schools, and a general approach to unity from the notion of oneness. In section 4, we will see why scientonomy seems to have pluralist elements, on the one hand, but on the other, unitarian elements.

3. The Plurality of Pluralisms

To begin with, we should say that there is not a single scientific pluralism. Instead, there are lots of different forms of pluralism based on different grounds, phenomena, and sciences. They vary in the strength, extension, and implications of the pluralism they advocate (Cat, 2017; Kellert, Longino, & Waters, 2006; Ruphy, 2017). The plurality of pluralisms is not just a fact, it is desirable. On the other hand, it has consequences for the natural limitation of this paper as it is impossible to talk about the various kinds of pluralism and go through their specific details and differences in a single paper of reasonable length.

To cover as much ground as possible, and to better cope with our limitations, our approach will focus on, first, clarifying pluralism in general terms. We will explore its relationship to the general idea of unity (oneness). Second, we will talk about schools or types of pluralism. This will help examine the relationship between pluralism and scientonomy.

The first thing to say is that pluralism, in which "multiple things go", should not be confused with relativism ("anything goes") or skepticism ("nothing goes"). In general terms, pluralism applies to concepts, disciplines, objects, values, explanations, models, methods, cultures, epistemologies, and ontologies. That diversity is, of course, a general feature of pluralism, and it does reflect its rejection of monism, essentialism, unity, uniformism, universalism, and reductionism (Cat, 2017).[2] According to Kellert, Longino, and Waters (2006), it is also useful to distinguish between pluralism *about* the sciences, and pluralism *in the* sciences: "a feature of the present state of inquiry in a number of areas of scientific research… [is that they] are characterized by multiple approaches, each revealing different facets of a phenomenon", including explanatory, modeling, and a theoretical plurality (Kellert, Longino, & Waters, 2006, p. ix). It is a view that "plurality in science possibly represents an ineliminable character of scientific inquiry and knowledge" (Kellert, Longino, & Waters, 2006, p. ix). That is useful because it helps us see where, how far, and in what direction pluralists take their proposals. *Pluralism about the sciences*, as we can see, is a simple position that there is not this single thing called science, but many (Kellert, Longino, & Waters, 2006, p. ix). As Ruphy reminds us, this last form of pluralism about the sciences is not up for debate. The fact that the sciences have branched out into numerous specialized disciplines is a fact assumed by both pluralism and unitarianism (Ruphy, 2017, p. xii).

Traditionally, pluralism is found in debate with monism.[3] Monism, which generally is taken to be an ontological approach, also has epistemic and methodological aspects. It could be fundamentally understood in terms of the attribute of oneness: the nature, fact, or state of being one or united. It could refer to one theory, one set of theories, one world view, one method, one object, one mosaic, one unit. The oneness can also be related to a *target* (the thing to which monism attributes oneness), and also to the *unit* (used to count something) (Schaffer, 2018, sec 1.1). According to Schaffer (2018), the strictest form of monism occurs when for a target T, counted by the unit U, the result is

---

[2] Some of these terms may be seen as overlapping and their meanings may vary from author to author.

[3] Since monism can be seen from an epistemic, methodological and ontological viewpoint, from we will not make specific separations in its use throughout this paper.

On the other side, the Minnesota school holds that its form of pluralism "is not based on metaphysical assumptions" (Kellert, Longino, & Waters, 2006, p. xiii). To this school, pluralism is a result of an empirically based interpretation, i.e., they avoid any *a priori* rationale to assess the natural world. So instead of saying, such as monists usually do (being fundamentalists or nonfundamentalists), that the natural world can be assessed from a single best account, they assume that this is "an open empirical question" (Kellert, Longino, & Waters, 2006, p. xi). Helen Longino's pluralism, as a member of this school, considers "the complexity of the world to champion the epistemic acceptability of the existence of several incompatible representations of a given phenomenon, claiming that the integration of these partial representations cannot be expected" (Ruphy, 2017, p. xiii).

There are other types of pluralism as well. For instance, Jordi Cat (2017, section 5.2) mentions four brands of pluralism, and how they dialogue with one another. Those four brands are vertical and horizontal pluralism, global and local pluralism, isolationist and integrative pluralism, and internal and external pluralism. *Vertical pluralism* is a form of factual description of inter-level pluralisms because each kind of fact is ontologically/conceptually autonomous, fundamental, and irreducible. *Horizontal pluralism* is the view that facts belonging to the same level are incompatible. *Global pluralism* is about every type of fact or description and *local* just about one. However, we can also have *global vertical pluralism* where there is no reduction of fact or description to any other, and local *vertical or horizontal pluralism* where just one type of fact or description happens on local intra-level or inter-level. *Isolationist pluralism* is about the choice from the disjunction of equivalent options, i.e., it is about underdetermination (Cat, 2017, section 5.2).

These brands of pluralism can be combined into another brand like global vertical pluralism or isolationist vertical pluralism. The reason for that is because vertical and horizontal pluralisms articulate the intra-level and inter-level relationship of the items inside any chosen form of pluralism, and that opens possibilities for dialogue.

The divergence of pluralisms may help us in our final task in this paper. For instance, the *integrative pluralism* assumed by Mitchell (2002), and which according to Kellert, Longino, and Waters (2006) is a form of moderate pluralism,[4]

---

[4] The term moderate does not mean necessarily a commonplace when most pluralists feel comfortable. According to Kellert, Longino, and Waters (2006, p. xi), the term moderate is related with the level of commitment to pluralism, i.e., it is about pluralists that assume pluralism but in principle they treat it as resolvable. This kind of pluralism is opposed to those that do believe that pluralism is ineliminable, such as the representational incompatible pluralism found in Longino (2006, p. 103) which, because of this aspect is considered radical pluralist by Ruphy (2017, p. 89).

is the conjunctive or holistic requirement of different types of descriptions and facts.

Ruphy (2017) says that integrative pluralism can also be seen as a form of *compatible pluralism* because the plurality of idealized causal models is "integrated to account for a particular concrete phenomenon" (2017, p. 86). That is interesting because while integrative pluralism acknowledges the plurality of models, it also assumes that different idealized causal models could cover different aspects of a single phenomenon, which must be integrated into a common framework for the understanding of that particular concrete phenomenon. So a compatible pluralism could be seen as an effort to integrate a plurality of descriptions about that phenomenon to offer a better representation of it.

The notion of representation, says Ruphy (2017), comes in several forms, such as theories, laws, computer simulations, taxonomic systems, models, and explanatory mechanisms. In general, representation aspires to provide a picture of a given phenomenon and relating it with truth, fit, conformation, similarity, correspondence, alignment, and so forth, depending on which school, and philosopher we are talking about. For instance, Longino assumes that we cannot have truth and precise explanations at the same time (Longino, 1990). While she rejects truth, at least as correspondence, to get better explanations of natural phenomena, the traditional analysis assumes truth as correspondence. Another case is Ronald Giere (2006), who like Longino (1990) rejects the traditional analysis of correspondence, but embraces representation as a relationship between models and the world under an idea of "successful fit" (Ruphy, 2017, p. 83), which can only be evaluated under its own perspective.

On this account, *compatible pluralism*, by and large, is the idea that despite a variety of explanations for a given phenomenon or different parts or features of it, these explanations are not mutually incompatible. In a more tolerant version, they may be temporarily incompatible, because researchers are seeking to resolve the incompatibility, and turn them into a common framework that they do not yet have (Ruphy, 2017). A methodological consequence of this, says Ruphy (2017), is the "epistemologically inoffensive" plurality of incompatible alternatives (models, explanations, computational systems, theories) facing the standard one (2017, p. 86). The inoffensiveness comes from the ontological status assumed for those alternatives, i.e., the alternatives are seen only as tools or tentative accounts warranting investigation, rather than reliable tools of representation (2017, p. 86).

Thus, inoffensive incompatible alternatives are, in reality, viewed only as tools in compatible pluralism. Besides, what is called *compatible pluralism* could be seen not only from the methodological perspective, but also from the metaphysical and epistemic perspectives, since such *a priori* inoffensiveness could not be sustained from a solely methodological outlook.

The relevant question for our aim is what we can do about such a plurality? Should we accept it from a methodological point of view but not as an epistemic and ontological approach? What do scientists do about it? Furthermore, what to do with such incompatibility of accounts, temporary or not, when contrasted with the notion of a scientific mosaic and the scientonomic requirements of the zeroth law? All of those forms of pluralism open different avenues to evaluate in what sense scientonomy can be seen as pluralist and also why some forms of pluralism are more consistent with it, especially with regard to the compatibility mechanism. On this account, we are going to see below how and why the zeroth law is pluralist.

4. Compatibility Through the Lens of the Pluralism of Compatible Theories and the Pluralism of Incompatible Mosaics

Now we are going to explore some doors opened in section 2 in light of section 3. Among the questions we raised in section 2, the chief one is about the nature of compatibilism in the zeroth law. This central issue asks if compatibility is an outcome of pluralism or monism in scientonomy's ontology, and in what sense. For instance, scientonomy admits the possibility of inconsistency between two or more accepted theories due to the complexity and fallibility of our approaches to natural phenomena through our limited theories. However, that inconsistency is taken as a temporary aspect of our knowledge, that is, an indication that our knowledge about some phenomenon at any given time is incomplete, limited, and fallible (Barseghyan, 2015).[5] It was not taken to be part of the complexity of the natural phenomena itself. Regarding that, for instance, Ruphy (2017) reminds us that at the core of the Vienna Circle unity of science program, there was a question which she thinks is still relevant, about "whether or not there exist different kinds of things that can be known only in different ways" (Ruphy, 2017, p. xv). This is the question of whether plurality is necessary to explore some complex natural phenomena. That is a relevant question for us, since scientonomy wants, by its laws, to describe scientific change.

Thereafter, the scientists' supposed despisal of inconsistency would lead them to undertake a quest to end such inconsistency. They would always be looking "into the future for the resolution to a single true account" (Sandra D Mitchell & Dietrich, 2006, p. S76). One could take that as part of a form of inter-

---

[5] It is important to highlight that some forms of pluralist debate are not very relevant to discussing the acceptance of theories or the worth of pursuing them. Scientonomy, correctly, does not involve itself in such debates.

theoretic reduction, manifesting itself in the mosaic as compatibility, i.e., like a form of monism. As Mitchell and Dietrich (2006) highlight elsewhere, Kitcher (1990) conceives of pluralism as a way of looking at a variety of alternatives due to epistemic fallibilism, aimed eventually "at universal acceptance of the true theory" (1990, p. 19). Thus, on the one hand, it would be possible to say that there is a monist view driving scientonomy's approach to scientific change. On the other hand, it would also be possible to argue for a pluralist viewpoint because, until that reduction takes place, scientonomy can deal with competing inconsistent accounts in a single mosaic. As we will see, it can also deal with incompatible accepted accounts outside a single mosaic model.

On the mechanism of compatibility, it is possible to assert that scientonomy describes science in a pluralistic way. However, as we have seen, not all forms of pluralism would fit well. Our task here is to argue about what kind of pluralism we are talking about. Therefore, an appropriate form of pluralism seems to be compatibilist pluralism for the reasons mentioned earlier. Amongst those reasons, for instance, while scientonomy posits monism inside the mosaic, it also tolerates a plurality of inconsistent accepted theories, even if only as a unified (fallible) accepted account concerning a phenomenon. That being the case, theories will compete to arrive at universal acceptance of a single account (or a set of compatible accounts) within the mosaic of its respective domain. This is reflected in the zeroth law, which does not tolerate incompatible accepted theories about the same phenomenon or domain coexisting within the same mosaic.

To make this claim clear, it is useful to remember some of the key features of the scientonomic mechanism of compatibility noted in section 2 (Barseghyan, 2015).[6] They are (1a) contemporaneous conflict about the same phenomenon from theories which belong to different domains and (1b) anomaly-tolerance between accepted general theory and a singular proposition within the same mosaic, about the same phenomenon.[7]

Those aspects of the mechanism raise the possibility that the zeroth law is pluralist, but based on compatibilist pluralism. Scientonomy focuses primarily on problems with respect to a single mosaic. For instance, case 1a is motivated by QT and GR each being seen as the best available description of their respective domains, but at the same time may be seen as competitors in the

---

[6] There are also additional two aspects of the mechanism of compatibility that we will not on touch here: 2) compatibility of non-exhaustive methods; 3) indirect incompatibility between theory and method, since they cannot be directly compared.

[7] Beyond the items '1a' and '1b' above, later on we will also look into the case of the *theory rejection theorem*, and the case of the *mosaic split theorem*.

search for a compatible situation within a single mosaic. Conversely, case 1b takes a single mosaic as its starting and finishing point.

Furthermore, we will see that a striking thing also happens with the mosaic, for the zeroth law will need to find a way to account for two accepted *incompatible* theories that initially belong to the same mosaic without actually violating the law. That exception is called mosaic split (Barseghyan, 2015, pp. 202-216). Whatever the case, in the end, scientonomy needs a way to tolerate anomalies, incompatibilities, and inconsistencies until the community finds good reasons to favor one account. Briefly, the question of how science deals with multiple accepted theories presses the zeroth law to adopt solutions like, the *mosaic split theorem*, and the *theory rejection theorem*. However, we must stress that any categorization of the zeroth law cannot simply be fit into the single box of pluralism or monism. That is why we proposed a refined kind of pluralism called compatibilist pluralism, which exposes that categorizing the mechanism of compatibility has an inner complexity.

As an example, let us turn to the axioms and theorems of theories,[8] case 1b of compatible pluralism. According to it, theories have axioms, so when any theory is accepted, its axioms may be incompatible with the axioms of the previously accepted one. In this case, the axioms of the old theory must be removed from the mosaic (Barseghyan, 2015, p. 168). That is a basic rule of the mechanism of compatibility. There is no way to leave two mutually incompatible axioms in the same mosaic, so when one theory is admitted, the old incompatible one has to be removed for otherwise "the mosaic would contain mutually incompatible elements, which is forbidden by *the law of compatibility*" (Barseghyan, 2015, p. 167). Besides that, with respect to theorems, we dissociate them from their rejected theories. So, whereas in the process of theory change the axioms of the old theory would need to be removed from the mosaic, this is not necessarily the case for the old theorems. That is called *the theory rejection theorem* (Barseghyan, 2015, pp. 168, 171). In short, the zeroth law can support a plurality of elements, especially theorems, in the mosaic at any time, regardless of whether the theorem in play belongs to the newly accepted theory or the old rejected one. Of course, there is a monist aspect to respecting the compatibility of axioms in mosaics, but also a compatible pluralism with respect to theorems.

What does that mean? When two theories are working to be the best available description of the *same domain*, but one of them is incompatible with the

---

[8] Here we define axioms as the structural elements of theories, i.e. their ontologies, because of the established relation between theory change and change in accepted ontologies, avoiding cooked-up explanations (Barseghyan, 2015, pp. 148-149).

other, then the axioms of both theories cannot be accepted in the same mosaic. In the contemporary scientific mosaic, this will happen when the new theory manages to assert and confirm novel predictions that are unpredicted from the old theory's viewpoint. This shows that the new theory has not cooked up explanations, and its new ontology will be approved (Barseghyan, 2015, pp. 149-150). To sum up, the old theory must be rejected, but its theorems need not necessarily be rejected. It is possible that the theorems of the old theory and those of the new theory will be compatible.

An example, given by scientonomists, is the theorem of *plenism* (i.e., there is no empty space), shared by the Aristotelian-medieval system and the Cartesian theory "for it was a theorem in the Cartesian system too" (Barseghyan, 2015, p. 169). Therefore, the compatibility of theorems, from two mutually inconsistent theories, opens the doors for a *plurality of sources* of theorems since incompatible theories can share compatible theorems. Of course, the mosaic still keeps a single accepted theory (or a set of compatible accepted theories) inside the same mosaic. Any form of competition is only seeking unification, or since we also have two or more compatible theories as accepted, we could say that it seeks compatibility of a plurality of theories inside the mosaic.

This is crucial since it also helps us to see that other forms of pluralism, such as ontological pluralism, conflict with scientonomy's compatible pluralism because the zeroth law respects the mechanism derived from theories inside the mosaic (we will notice it clearly in cases of mosaic split, where we will use the expression *pluralism of incompatible mosaics*). Let us now explore the well-known black hole example, scenario 1a, to consider this pluralistic debate.

By addressing that matter, the mechanism of compatibility rests on a distinction between compatibility and the notion of inconsistency. That is a key aspect of the mechanism of compatibility to which Harder (2013) drew our attention. He argued that the compatibility of two accepted theories is possible even when they are logically inconsistent because there are historical cases where we find that situation, and also because we cannot see or predict all logical consequences implicit in two compatible and inconsistent theories (Barseghyan, 2015, p. 153).

Combined, compatibility and inconsistency embrace the complexity of the world and quasi-true research results. Besides, they also acknowledge the practical issues that guide scientists to work with contradictions when they occur. Thus, let us start with the black hole example. If we look carefully at the case of QT and GR, we see that despite belonging to the physics mosaic, they explain different domains of phenomena (respectively micro and macro phenomena). Ordinarily, they do not conflict with one another, *except* when applying to such phenomena singularities within black holes. Both accepted theories belong to different domains, and in a certain sense, also to two different mosaics of two subcommunities of physicists. These mosaics are focused on several dimensions of natural phenomena, which may entail completely opposing ontologies reflected in their theorems and methods. That is why, according to the zeroth law, they can be accepted without any mosaic split, despite the fact that they are logically inconsistent: there would still be *one* best explanation in each distinct domain. In the case of singularities, the interests of GR and QT overlap, and this is when scientonomy explains the situation of those two theories from the compatibilist viewpoint.

When that happens, strictly speaking, QT and GR are still the only accepted views in their respective domains, so there is not necessarily pluralistic compatibility here. In other words, the ontological oneness of the mosaic (one set of compatible theories or a single theory) is not violated because QT and GR are dealing with sufficiently different classes of phenomena. A single phenomenon that is 'too complex' such as black hole singularities does not change the fact that QT and GR belong to strictly different areas of physics interested in different sets of phenomena.

In that sense, scientonomy and the mechanism of compatibility are, from one viewpoint, monistic because the law of compatibility forbids mutually incompatible elements in the same mosaic, despite its tolerance for logical inconsistencies. However, things are not this simple, or as Ruphy said, "it does not make much sense to claim to be a monist or a pluralist tout court about science today" (Ruphy, 2017, p. xiv). As we saw in section 3, things start to get complicated when two accepted theories provide inconsistent descriptions of the same phenomenon. However, the problem is that cases such as the black hole singularities are so complex that, in principle, they require both QT and GR. And most of the time, despite QT and GR providing mutually inconsistent accounts of singularities, they also do not enter into conflict with regard to many other phenomena. In the rare case when they do conflict, this produces a form of compatible pluralism.

What about when that happens? According to Barseghyan, in principle, both theories should in principle be applicable to the phenomenon of singularity (Barseghyan, 2015, p. 159). Nevertheless, they cannot be both simultaneously applied to the phenomenon due to their apparent logical inconsistencies. Rather "the community seems to be limiting the application of at least one of the two theories" (Barseghyan, 2015, p. 159). Insofar as the scientific community restricts the application of one of those theories, "we accept the two theories only with a special "patch" that *temporarily* limits their applicability" (Barseghyan, 2015, p. 159).

Nonetheless, scientonomy still argues that both theories, despite their inconsistencies evidenced by such unusual cases, remain accepted in the mosaic. On this account, the zeroth law could be studied from a synchronic view (a static perspective with respect to changes in the mosaic). As a consequence, QT and GR are part of the same mosaic, although when talking about singularities they need to use paraconsistent logic. This is necessary even to decide that they should both remain accepted in the mosaic with guidelines for the sake of establishing a 'patch' that limits the applicability of both theories, or at least of one of them.

On the other hand, we know that despite both theories being focused on their specific domains (micro and macro), this does not mean that these are the only domain for which they have implications. The two theories are also interested in extending their explanations over other domains. Thus, sometimes cases like that of singularity provide an opportunity for competition between two or more theories. Due to the fact that scientists actively try to eliminate contradictions, that competition must result in some mosaic unification or mosaic subsumption. In this case, the intended goal is the unification of the two theories, resulting in a logically consistent theory of everything. On this account, while we can say that there is some pluralistic tolerance in the QT and GR competition, as part of a particular temporary patch, we can also say that temporary tolerance of inconsistency between two accepted theories will eventually be terminated, or at least that is what scientonomy seems to expect.

Thereby, maybe scientonomy considers its expectation of the elimination of inconsistency to result from some form of intertheoretical reducibility, implicit in the zeroth law, which "amounts to the possibility that one theory or area of discourse is absorbed or subsumed into another" (Ruphy, 2017, p. xvii). That form of compatibilism develops plurality by aiming at different types of entities and relations, or as Kitcher (2001) puts, different *intended content*. According to Ruphy (2017), this is an example of compatible pluralism because of the metaphysical belief in some form of ordered world reflected in the constant attempts to "put together [all available views] to provide a global picture of the phenomenon" (Ruphy, 2017, p. 86, brackets added). To offer a contrasting view, coming from pluralists such as Longino (1990, 2002, 2006) or Dupré (1993), it is said that there can be options such as acknowledging the non-integrable nature of the phenomenon and the incompatibility of our theoretical frameworks. In other words, "adopting only one model (and rejecting the others) is not the implicit goal" (Ruphy, 2017, p. 86). It is not even seen as an implicit goal, at least if we want to avoid dogmatic metaphysics. As Kellert, Longino, and Waters (2006) said, it should be seen as an open empirical question.

Let us now turn to the scenario 1b again, with a focus on the issue of anomalies. An anomaly occurs when an accepted theory conflicts with a singular proposition concerning the same phenomenon rather than with another theory. In a nutshell, since in the anomaly-tolerance scenario, we could have one or more anomalies, there is a kind of pluralistic approach in the mechanism of compatibility. On the other hand, there is a kind of unity in the mosaic since the accepted theory remains accepted. The existence of the anomaly of course does not imply that there is any other incompatible theory within the mosaic that could serve as an alternative to the accepted theory.

Now, suppose that a new theory is accepted that explains the anomaly while the previously accepted theory also manages to keep its place in the mosaic. In such a scenario, we have a manifest dispute between accepted incompatible worldviews. How does scientonomy deal with this situation from the point of view of the mosaic and the mechanism of compatibility? That is when, we argue, scientonomy turns from a *theoretical compatibilist plurality* of single mosaics into a *pluralism of incompatible mosaics* by invoking the concept of mosaic split.[9]

A mosaic split takes place when an "initially united community" with a single mosaic for theories and methods transforms, due to a peculiar form of disagreement about theories (on their very status of acceptance or not), into two or more communities with two or more mosaics (Barseghyan, 2015, p. 202). So "as long as the members of two groups have the same opinion on what the accepted theory is, the scientific community remains united" (Barseghyan, 2015, p. 202). When they disagree about that, the mosaic may split.

As is well known, the zeroth law can tolerate, within the same mosaic, a plurality of theories as accepted, but just if, and only if, they are mutually compatible (regardless of any logical inconsistency). Thus, a mosaic split occurs when scientific investigation motivates a scientific community to split into two communities accepting two different theories, which are mutually incompatible concerning the same domain. Hence, scientonomy's solution for cases like these is to split the mosaic into two or more mosaics until a solution for the disagreement prompting the split is discovered. Theoretically, plurality can only occur *within* a mosaic of compatible theories. However, outside that single mosaic, there is no theoretical pluralism, seen only from an epistemological approach. Nonetheless, the mosaic split pushes forward a plurality of ontologies, which reflects onto a *plurality of mosaics with incompatible and accepted theories.*

---

[9] Mosaic split can take place under two scenarios: *the necessary mosaic split theorem* and *the possible mosaic split theorem.* We will deal specifically with the instance of actual mosaic split, so it does not matter if its origin is *the necessary theorem* or the *possible theorem.* It is important to see how cases of mosaic split help us to understand this aspect of plurality in scientonomy.

The reason is that more than one theory may fulfill requirements of currently employed methods, and be accepted and incompatible (not just inconsistent) with the still accepted theory. As we quoted elsewhere, the zeroth law cannot tolerate that kind of plurality. So considering that scientonomy sees achievement or maintenance of a single mosaic as a constant aim of any scientific community, an implicit goal, the outcome of such apparent contradiction is the splitting of the mosaic into two or more mosaics.

The mechanism of compatibility cannot escape that consequence, especially because when it relates with the other laws, such as the second law, it tells us that if a theory meets the requirements employed at the time, the result is that we should accept it even if the new theory is incompatible with the already accepted one (Barseghyan, 2015, p. 204). That is ordinarily when the requirements of the methods at the time, as the Aristotelian requirement of "intuition schooled by experience", are too not overly restrictive, resulting in cases of mutually incompatible and accepted theories, such as the acceptance of the Cartesian and Newtonian views in the 18th century.

Considering that each mosaic can bear only mutually compatible theories at any one time, scientonomy's way out for cases where two or more incompatible theories are accepted is splitting the mosaic because "we are left with only one logically available option" (Barseghyan, 2015, pp. 203-204). In a word, instead of just being pluralist, respect accepted and compatible theories within a single mosaic, what scientonomy does, at least temporarily, is to ask us to be pluralist with respect to mosaics, constituted by accepted and incompatible theories.

Therefore, the pluralistic aspect in scientonomy is a pluralism of compatible theories and pluralism of incompatible mosaics. It is of particular interest to say that pluralism of incompatible mosaics multiplies the number of mosaics, and in each mosaic, there would be a successful scientific theory, entirely different in content.

Although *pluralism of incompatible mosaics* is not explicitly formulated by Barseghyan (2015) or by scientonomists' papers related to the mechanism of compatibility, it does not seem to be incompatible with their view of scientific change. Moreover, the competition of theories within different mosaics dealing with the same domain of science, brings up an ontological challenge for scientonomy. For instance, the competition in the scientific revolution between Aristotelian, Cartesian, and Newtonian worldviews does not only reflect a theoretical competition but since it is a case of a mosaic split, there is also a change in ontologies. Thus, the pluralism of mosaics seems to be helpful in this kind of competition because the history of science sometimes exposes profound differences between two theories, which go beyond just epistemic or methodological disputes.

5. Final Considerations

The discussion of compatible and incompatible theories and its implications for the mechanism of compatibility and the mosaic has only been superficially scratched. Nonetheless, the arguments presented above leave no doubt that scientonomy's understanding of compatibility is pluralist for theories inside the mosaic. Moreover, our discussion raises questions about which kind of pluralism is better suited to scientonomy's purposes. One of the reasons for such a challenge is how scientonomy describes scientific communities and how their approach to the variety of accepted theories with different ontologies also seems to include some monist aspects. In this account, to fulfill that challenge, we argued that scientonomy's compatibilism fits better with compatible pluralism, at least if we are addressing theories that are trying to change or maintain their status in the mosaic.

Furthermore, Scientonomy also talks about cases of mosaic split, when more than one theory is accepted in the same mosaic, but its split is needed because those theories are not compatible inside the mosaic. That is when scientonomy needs to use the *pluralism of incompatible mosaics* in order to describe better cases of history of science where two or more theories are accepted, but due to the laws of scientific change, are unable to occupy the same mosaic. Therefore, scientonomy seems to be pluralist, at least under those two forms, and that works well in its descriptive task to understand how science works dealing with some complex natural phenomena.

Bibliography

Barseghyan, H. (2015). *The Laws of Scientific Change*. Springer.

Barseghyan, H. (2018). Redrafting the Ontology of Scientific Change. *Scientonomy*, 2, 13-38.

Cartwright, N. (1999). *The Dappled World: A Study of the Boundaries of Science*. Cambridge University Press.

Cat, J. (2017). The Unity of Science. In Zalta, E. N. (Ed.) (2017). *The Stanford Encyclopedia of Philosophy (Fall 2017 Edition)*. Retrieved from: https://plato.stanford.edu/archives/fall2017/entries/scientific-unity/.

Chang, H. (2012). *Is Water H_2O? Evidence, Realism and Pluralism*. Springer.

Dupré, J. (1993). *The Disorder of Things: Metaphysical Foundations of the Disunity of Science*. Harvard University Press.

Ereshefsky, M. (2017). Species. In Zalta, E. N. (Ed.) (2017). *The Stanford Encyclopedia of Philosophy (Fall 2017 Edition)*. Retrieved from: https://plato.stanford.edu/archives/fall2017/entries/species.

Giere, R. (2006). *Scientific Perspectivism*. University of Chicago Press.

Harder, R. (2013). *Scientific Mosaics and the Law of Consistency*. Unpublished Manuscript.

Kellert, S. H., Longino, H. E., & Waters, C. K. (2006). Introduction: The Pluralist Stance. In Kellert, Longino, & Waters (Eds.) (2006), vii-xxix.

Kellert, S. H., Longino, H. E., & Waters, C. K. (Eds.) (2006). *Scientific Pluralism*. University of Minnesota Press.

Kitcher, P. (1990). The Division of Cognitive Labor. *The Journal of Philosophy*, 87(1), 5-22.

Kitcher, P. (2001). *Science, Truth, and Democracy*. Oxford University Press.

Longino, H. (1990). *Science as Social Knowledge: Values and Objectivity in Scientific Inquiry*. Princeton University Press.

Longino, H. (2002). *The Fate of Knowledge*. Princeton University Press.

Longino, H. (2006). Theoretical Pluralism and the Scientific Study of Behavior. In Kellert, Longino, & Waters (Eds.) (2006), 102-131.

Mitchell, S. D. (2002). Integrative Pluralism. *Biology and Philosophy*, 17, 55-70.

Mitchell, S. D. & Dietrich, M. R. (2006). Integration without Unification: An Argument for Pluralism in the Biological Sciences. *The American Naturalist*, 168(S6), S73-S79.

Morrison, M. (2011). One Phenomenon, Many Models: Inconsistency and Complementarity. *Studies in History and Philosophy of Science Part A*, 42(2), 342-351.

Patton, P., Overgaard, N., & Barseghyan, H. (2017). Reformulating the Second Law. *Scientonomy*, 1, 29-39.

Ruphy, S. (2017). *Scientific Pluralism Reconsidered: A New Approach to the (Dis)unity of Science*. University of Pittsburgh Press.

Schaffer, J. (2018). Monism. In Zalta, E. N. (Ed.) (2018). *The Stanford Encyclopedia of Philosophy (Winter 2018 Edition)*. Retrieved from: https://plato.stanford.edu/archives/win2018/entries/monism.

Sebastien, Z. (2016). The Status of Normative Propositions in the Theory of Scientific Change. *Scientonomy*, 1, 1-9.

Suppes, P. (1978). The Plurality of Science. *PSA: Proceedings of the Biennial Meeting of the Philosophy of Science Association*, 3-16.

Chapter 11

Historical Advances in Ecology

Jamie Shaw

University of Toronto

Justin Donhauser

Bowling Green State University

Abstract: The purpose of this paper is to show that historical advances in theoretical ecology do not conform to the conventional scientonomic ontology. As a result, we suggest some revisions of that ontology. We claim that three famous episodes, including the development of three kinds of models (Lotka-Volterra models, broken stick models, and exergy models), demonstrate the need for a few modifications. Specifically, they highlight the need to refine the scientonomic category of *use* into two distinct kinds: *epistemic use* and *practical use*. Moreover, we suggest introducing the notion of *abstract theory*. We go on to argue that these historical findings support the recent changes in the definition of *theory acceptance*.

Keywords: Lotka-Volterra Models; epistemic use; acceptance; models

***

1. Introduction

The purpose of this paper is to show that historical advances in theoretical ecology do not conform to the conventional ontology in scientonomy and suggest revisions. We claim that three famous episodes, including the development of three kinds of models (Lotka-Volterra models, broken stick models, and exergy models), demonstrate the need for a few modifications of the accepted ontology. Specifically, it highlights the need to refine the scientonomic category of *use* into two distinct kinds: *epistemic use* and *practical use*. Moreover, we suggest introducing the notion of *abstract theory*.

We go on to argue that these historical findings support the recent changes in the definition of *theory acceptance*.

In a previous variation of this paper, we argued that some of the most landmark advances in theoretical ecology were consistent with voluntarism, or the view that our development of theoretical apparatuses is largely determined by free choices in accordance with particular goals (Donhauser & Shaw, 2019). In other words, they were consistent with a kind of voluntarism where models were transferred across domains in accordance with a wide range of idiosyncratic epistemic needs. Moreover, we argued that these case studies demonstrated serious flaws in arguments from incommensurability. Arguments from incommensurability, especially of a strong neo-Kantian flavor, suggest that transfer of these models should be difficult, if not impossible, given that the communities live in "different worlds". But the lessons that can be garnered from these historical episodes have not been exhausted quite yet. Specifically, in this paper, we demonstrate that a proper appreciation of these case studies suggests some refinements and revisions to the framework currently accepted within scientonomy. We contend that the current ontology of epistemic stances and elements requires adapting to accommodate a proper reconstruction of these episodes in 20th-century ecology.

There are two primary halves to this paper. In the first half, we are recapitulating the case studies on the Lotka-Vottera, exergy, and broken stick models. After this, we show how these historical case studies recommend a few revisions to the extant scientonomic framework. Specifically, we argue the need to distinguish between different kinds of *use* and different kinds of *theory* that epistemic agents can take stances towards. We also mention how these case studies reinforce the newest understanding of *acceptance* over its predecessor.

2. Lotka-Volterra Models

As Donhauser & Shaw (2019) explain, the development of Lotka-Volterra models (LV-models) was incredibly important. This is partially due to the fact that they can be used to characterize vastly different sorts of phenomena throughout ecology and many other sciences. LV-models use basic ideas about rates of reciprocal interactions initially developed in chemical kinetics (after Lotka, 1910) to characterize, and thereby aid in understanding, an array of vastly different phenomena. LV-models have been a staple of many facets of ecological research since their introduction in the early 1900s (see Codling & Dumbrell, 2012, p. 144). In many applications in ecology, LV-models characterize dynamics between biological populations interacting as predator and prey with equations initially developed used to characterize reciprocal dynamics between different kinds of chemicals.

LV models, in ecology, are composed variables for species-typical efficiencies, relationships, and mortality rates. Their values and mechanisms of realization change depending on the populations one selects to play predator and prey roles in her model community. A simple LV-model can be generated by filling in the values of two differential equations:

1) $\frac{dD}{dt} = aD - bDW$

2) $\frac{dW}{dt} = -cW + dDW$

where D is the number of members of a simulated prey population, W is the number of members of a simulated predator population for that prey population, t is the elapsed time since the beginning of the simulation period [T], a is the growth rate of the prey [1/T], b is the parameter that quantifies the effect of predators on prey mortality, *c* is the death rate of the predator [1/T], and d is the parameter that quantifies the effect of prey consumption on the growth of the predator (Donhauser, 2017, p. 73).

LV-models generated with variations of these simple equations enable one to estimate the change in a selected prey population's abundance across time [dD/dt] by assuming that it is jointly determined by that population's species-typical growth rate and the efficiency of the predator population [aD − bDW]. Such models also enable one to estimate the change in the abundance of that predator population across time [dW/dt] by assuming that this is jointly determined by its species-typical mortality rate and the species-typical rate at which that predator population increases in abundance as a result of consuming prey (that population's 'conversion efficiency') [−cW + dDW].[1]

Researchers use LV-models to simulate population dynamics that should result from species-typical feeding behaviors and mortality rates *if* the assumptions of the model are accepted for target populations (cf. Odenbaugh & Alexandrova, 2011). As such, models take account of several variables and draw them together, and apply LV-model variations to simulate how many very different sorts of populations may change relative to one another according to means acceptable to the community of the time. In sketch, this is done by identifying natural populations that can be reasonably seen as interacting as predator and prey and then inputting population-specific values for the above-listed variables (e.g., species-typical birth and death rates).

---

[1] Lotka's (1925, pp. 92-93) original discussion of these equations is exceptionally clear and concise, and there he explains some of the rationale behind the construction of these equations.

As Donhauser has explained elsewhere (Donhauser, 2017), natural predator and prey populations can be located by a very simple criterion: for any two species, whether one is effectively a predator or prey for the other is just a matter of *whom kills whom.* Predators do the killing and prey are their targets. For instance, for dragonflies and bees, dragonflies are the predator because they eat bees and not the other way around. At the same time, dragonflies are potential prey for many species of frogs, since frogs eat dragonflies and not the other way around. With this simple whom-kills-whom criterion, "one can determine on the basis of observations or prior knowledge of species-typical behaviors which natural populations are predators and which are prey" (Donhauser, 2017, p. 73).

On what grounds can such a model be generalized to different types of interacting substances and populations? This is a particularly acute question for ecologists, because, let alone using models of chemicals to characterize organisms, individual members of the types of entities they investigate are indeed "complex, inherently variable, and functionally diverse" (rear cover, Reiners & Lockwood, 2010). Due to the high degree of variability within ecological types, it is often unclear whether successfully simulating some selected instances of an ecological type (e.g., a predator population) with a model shows that similar success can be expected when simulating other instances.

For example, using an LV-model to accurately predict paramecium (predator) and yeast (prey) population dynamics – the purpose for which the model was initially developed – does not show that such a model should also be able to predict wolf and deer population dynamics as accurately as it does (see Jost et al., 2005). Of course, this is because paramecium predation and wolf predation are extremely different and the relevant sorts of populations are comprised of very different species (cf. Peters, 1991, p. 57). Yet, even when the selected instances of an ecological type being modeled are not drastically different, the issue remains because population-level behaviors vary greatly both within and across species and because particular populations often exhibit species-atypical behaviors due to *in situ* environmental factors. So, even though it is reasonable to infer that an LV-model will aptly characterize populations of a species that that model has been successfully used to simulate in the past, this cannot be guaranteed and population-specific contingencies often need to be added in to approximate to observed dynamics with LV-model simulations (cf. Turchin, 2001; Donhauser, 2017). Of course, this still leaves open the question of what grounds there are for thinking that LV-models could be used to usefully characterize substantively different sorts of phenomena.

We believe the answer to this question is that the ways of representing and conceiving of interacting substances and entities when generating LV-models are phenomena neutral. Variations of the equations used to generate such models, (1) and (2), are not about anything in nature themselves. LV-models are models of selected targets of which such models are generated by filling in target-specific values; such that target-isomorphic (e.g., substance-isomorphic or population-isomorphic) differences create different LV-models.[2] Even if we focus on their applications in ecology alone, we see that the building blocks of LV-models are phenomena neutral since the 'predator' and prey' categories used to map the LV-equations onto natural populations are phenomena neutral and admit applications over a very broad range of interacting populations and substances.

We have mentioned varied interacting species, ranging from wolves, to bugs, to paramecium. But that listing does not even begin to broach the variety of relationships that can be simulated with LV-models. Indeed, the ecological literature shows exceedingly many applications that extend well beyond what one might naturally conjure when initially thinking about predator/prey populations. This is because the 'whom kills whom' criteria itself is actually more general, and something more like 'what impacts the mortality of what'. Accordingly, variations of LV-models are extended to characterize populations interacting through interspecies competition (-/-), predator/prey (+/-), mutualistic (+/+), amensalistic (-/0), and commensalistic (+/0) interactions – among others. Donhauser (2017) points out that they are also regularly extended to characterize reciprocal dynamics between plant populations in which one species (the 'predator') typically outcompetes the other (the 'prey') for light and nutrients (see, e.g., Mäkelä & Hari, 1984). Such models are even used in certain domains to characterize the reciprocal dynamics between diseases (the 'predator') and different species of organisms (the 'prey') (see, e.g., Holt & Pickering, 1985) and to make inferences about the impacts of changing chemical variables on population dynamics (see, e.g., Tett & Wilson, 2000; Thornley, Bergelson, & Parsons, 1995). Further still, it is instructive to note that more complex LV-models can be generated to explore seemingly innumerable possibilities regarding particular populations by simply combining more basic LV-models of these various sorts.

Simple LV-models conceptually connect two interacting populations (as in Figure 11.1) via the variables expressed in equations (1) and (2) above.

---

[2] Another way to think about phenomena-neutrality is to say that models and theories needn't entail commitments about the world but serve as 'ways of seeing things' or 'stances' that can be mapped onto different metaphysical pictures (cf. van Fraassen, 2008, pp. 246-250).

Figure 11.1: Conceptual diagram of the reciprocal causal impacts represented as causally connecting target wolf and deer populations in a basic LV-model (adapted from Donhauser & Shaw, 2019).

Generating LV-models to simulate more complex possible scenarios just entails linking together simple LV-models to conceptually map and connect more potential interacting populations and other quantities.[3] Figure 11.2 is offered to help one imagine how such models are generated (see also Donhauser, 2014, pp. 101-102).

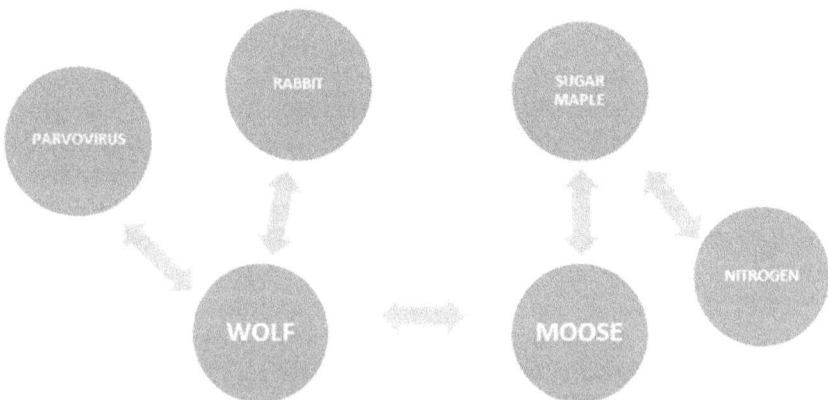

Figure 11.2: Conceptual diagram of the reciprocal causal impacts that can be represented as causally connecting other study-relevant populations and environmental factors to target wolf and deer populations in a more complex LV-model (adapted from Donhauser & Shaw, 2019)

The possible model suggested in Figure 11.2 considers organic populations and substances not conventionally conceived as predators or prey, but which can be usefully modelled as such because of the phenomena-neutrality of LV-modeling methods.

---

[3] Many sources explain specific methods of doing this; Bender, Case, & Gilpin, 1984, is exemplary.

3. Exergy Models

Next, we consider a general theory developed to understand phenomena in one domain that has been applied to re-envisage, and push forward understanding of, phenomena investigated in another. Many such examples exist in ecology and its development via applications of principles and models from general systems theory. Here we will focus on one such transfer through the uses of 'Exergy Theory' for simulating ecosystem dynamics.[4]

Exergy Theory is a thermodynamic theory – influenced by efforts to expand on Carnot's theory of heat in the mid-1800's – according to which the component parts of any thermodynamic system will tend to interact in ways that 'maximize energy available for work within that system' or 'exergy' (Rant, 1956; Susani et al., 2006). More precisely, Exergy Theory assumes that thermodynamic systems dissipate energy toward stasis, in accordance with the second law of thermodynamics, and that any such system's nodes at once exhibit dynamical interrelations such that the whole network of them counteracts that dissipation to the extent permitted by energy inputs to that system. Ecologist Sven Jørgensen calls this core principle of Exergy Theory the "tentative fourth law" to reflect that it augments accepted thermodynamic laws if it holds (Jørgensen, 2000) – we will call it the 'MaxEx principle'.

Ecologists who embrace MaxEx see ecosystems as thermodynamic networks. They define the organisms as nodes in those networks in thermodynamic terms, and, for certain purposes, think of them as being causally connected via their functional roles as 'vectors for the storage and transference of potential energy for work' within their networks. Those working with this view submit, moreover, that ecosystems are ecological networks that are disposed to move away from thermodynamic stasis by manifesting overall network dynamics that maximize exergy within their boundaries (to the extent permitted by energy inputs). In this way, they propose that ecosystems can be usefully conceived as behaving according to MaxEx.

Ecologists working with these ideas have done several studies in which they assess the "quality" of selected natural ecological networks by estimating their respective "exergetic potential" – the amount of work possible to extract from a system at a time in principle. They do this by totaling the net quantities of everything within an ecological network that can be converted into usable energy in principle. These include, but are not limited to, net biomass, heat energy, and

---

[4] Another good example is G. Evelyn Hutchinson's (1948) application of Norbert Weiner's (1948) 'Cybernetic' theory to aid in understanding ecosystem and community level dynamics. This is discussed in Donhauser, 2016; but see also Keller, 2008, and Slack, 2011, pp. 236-237.

molar concentrations of the various chemicals within a selected target network (Mitsch & Jørgensen, 2004, p. 92). In many studies, ecologists working in this vein also simulate how selected 'natural ecosystems' behave and evolve as a result of changing environmental variables with model ecological networks that incorporate mathematical interpretations of MaxEx (see Zhang, Gurkan, & Jørgensen, 2010). They often simulate how both the exergetic potential and variable parameters of selected ecological networks can change due to changes in the amount of some source of energy input to them.

The salient models enable such simulations by using mathematical expressions of MaxEx to help characterize thermodynamic relationships that would make a thermodynamic system tend to maximize its overall exergetic potential at any time *if* those relationships were realized in nature. This is done with the following equation:

$$3) \quad E = R T \sum_{i=1}^{n} \left(c_i \ln\left(\frac{c_i}{c_{i,eq}}\right) + \left(c_i - c_{i,eq}\right) \right)$$

where R is the gas constant within the modelled network, T is the absolute temperature, i is some potential energy source (e.g., nitrogen), c_i is the concentration of component i, and $c_{i,eq}$ is the estimated concentration of i at thermodynamic equilibrium (see, for example, Coffaro, Bocci, & Bendoricchio, 1997, p. 104; Zhang, Gurkan, & Jørgensen, 2010, p. 695).

In these studies, (3) serves as a mathematical function that partially dictates the behavior of simulated ecological networks by stipulating that they always "develop towards a state with a higher value [relative to a selected] goal" (Müller & Leupelt, 1998, p. 8). Here the 'goal' metaphor describes the use of an equation, such as (3), as an algorithm that dictates the behavioral arc that a simulated ecological network will exhibit in response to different values given for that simulated network's variables. Because of this, a mathematical expression of MaxEx can be applied to define the goal function in simulated communities and ecosystems.

Taking as given that the MaxEx-principle is an acceptable characterization of the general arc of ecological network dynamics (i.e., a tendency towards internal exergy maximization), numerous ecologists argue that ecosystem models that incorporate it as goal function enable more realistically dynamic simulations of natural populations than models that do not. Some have substantiated this claim by showing that models that use variations of the MaxEx-principle to estimate parameter range values and vary them in response to changes in environmental variable values more accurately simulate natural community dynamics than models 'parameterized' by other conventional methods. Proponents of this approach endorse it as a complement to the traditional, data-driven, method of estimating parameter range values. And looking at how it complements the traditional method helps clarify how the MaxEx-principle is applied as a goal function to develop more realistically dynamic ecological models.

That traditional method of parameter range estimation consists of a three-step process whereby researchers find possible range values for as many parameters as they can, determine the sensitivity of the parameters, and then calibrate the most sensitive parameters to give the best accordance between observations and model predictions/simulations (cf. Jørgensen, 2001, p. 299). In contrast, using the MaxEx-principle to dictate a goal function skips straight to the final calibration step. It consists in estimating parameter ranges by calculating the parameter values that collectively produce the highest 'exergetic potential' under environmental conditions specified in a model. By using the MaxEx principle this way, one can estimate relative parameter ranges for modelled variables using fixed values for selected environmental input variables. One can also dynamically vary all of the parameter values in a simulated ecological network in unison to simulate how ecological network dynamics may change in response to changes in selected environmental inputs using MaxEx.

It is rather easy to see how the parameter ranges of individual variables within an ecological network can change in response to changing environmental inputs to that network. For example, consider how growth rate parameters for a deer population could change in response to changes in the amount of nitrogen available to the plants they typically eat. As an illustration, imagine that when regular levels of nitrogen are present the birth rate for a particular deer population is 40-50 offspring per year and when relatively higher levels of nitrogen are present, and there is therefore an increased plant food supply, that deer population's birth rate increases to 70-80 offspring per year. With a basic understanding of how parameter ranges can vary relative to changing inputs in this way, it is then a short step to seeing how the MaxEx can be applied to estimate parameter ranges for a small number of selected variables in a model.

This is done by calculating parameter values for variables in a simulated ecological network that would collectively produce a 'goal' state determined via calculations with mathematical expressions of MaxEx like (3). In other words, one uses such expressions of MaxEx to calculate the parameter values that would collectively produce the highest net internal usable energy for that entire simulated network under conditions specified within the model. Imagining how researchers could use this method of parameter range estimation to simulate the dynamics of complex ecological networks in response to changes in environmental conditions is another short step, as the procedure for doing this simply entails doing many more calculations. Specifically, it entails calculating the parameter values for all modeled variables that collectively produce the highest exergetic potential of a modeled network with each change in select environmental variables for any number of environmental variables.

There have been many successful applications of exergy models reported in ecology; showing that such models are useful for characterizing various phenomena from plants to fish to finches; among others (see Zhang & Jørgensen, 2010). We think these cases make the point made with Lotka-Volterra models even more forcefully. Because they don't only show that theoretical principles and models can be and are phenomena neutral, but also show that phenomena studied in different domains can be easily reimagined differently. For example, one can see wolves and deer and plants not just as organisms but as micro-thermodynamic systems that serve as vectors of energy storage and transfer in bigger thermodynamic networks. And again, it seems that the issues of incommensurability are nowhere to be seen in the scientific practices we are looking at here.

4. Broken Stick Models[5]

Finally, there is yet another interesting development in the history of ecology, which is highlighted by visiting another episode that comes in between LV-models and the popularization of Exergy Models. This is the debate about applications of Null models, which grew out of the development of Niche Apportionment models originally introduced by Robert McArthur (the academic grandson of G. E. Hutchinson). To be very clear, null models are interesting because they seem to directly challenge a presupposition of scientonomy. This is because models are like mini theories as they are understood in the current scientonomic literature; because models are essentially seen as expressible as sets of propositions. Yet, null models are not used in a way that relies on accepting propositions, but often rely on the modelling process itself to garner new insights. McArthur's "Broken Stick" models illuminate what we mean.

A B-stick model is a way of representing a community-level ecological network that enables one to generate predictions about the relative abundances of the populations that comprise a particular community. B-stick models assume that relative population abundances are determined by how much of a shared resource each population consumes, and that populations will "competitively exclude" each other from an area (impact each other's overall growth and mortality rates) through their consumption of shared environmental resources. B-stick models also assume that the relative

---

[5] The discussion in this section is adapted from Donhauser, 2015.

abundances of the populations comprising any simulated community can be predicted by stipulating that each simulated population inhabits a nonoverlapping continuous niche.

According to the conception of niches that MacArthur (1957) employs, niches are subdivisions of an area (e.g., a forest) that are uniquely occupied by each population in that area, and whose boundaries are delineated by the amount of resources each population uses. MacArthur does not say anything to indicate whether niches are to be understood in mereotopological terms (as spaces uniquely occupied by discrete populations) or in functional terms (as consumer roles played by discrete populations). In fact, he only says that, for the purpose of generating B-stick models, he conceives of niches as "continuous", in the sense that they are not spatially dispersed in patches throughout an area, and as "nonoverlapping" in the sense that each population's niche excludes those of other populations. Interestingly he admits that this conception is false, since populations do actually mereologically and functionally overlap, but then he demonstrates that this conception is more predictively powerful and efficient than data models of niches. In scientonomic terms, MacArthur's broken stick models are not meant to be accepted. What we think is more interesting and may have profound implications for scientonomy, is that Broken Stick models use only these wrong assumptions but are then generated through a process that determines their final structure. So the process is doing the work, not the assumed propositions.

Procedurally, following MacArthur's process, one generates a B-stick model of a selected natural target community or a possible community via the following five steps.[6]

1.   One imagines a line, or "stick", an arbitrary number of units long, to represent the total shared resources available to all of the populations being simulated. For simplicity, I will here make the number of available resource units 200 (Figure 11.3).

0   **200**

Figure 11.3: An Interpretation of a MacArthur Stick

---

[6] Figures 11.3-6 are adapted from Wilson, 1993, p. 182. These figures are purely illustrative and are *not* scaled to reflect the typical abundance schema of any sort of natural community.

2.   One then segments this stick so that there are segments corresponding to the number of distinct populations being simulated (*n*). This is done by making one less (*n*-1) random "throws" at the line – where what MacArthur calls "throws" are randomly assigned points on the line that are generated in a manner like throwing darts at random points on a line. Hence, for a five-species (*n*) community one makes four (*n*-1) throws (Figure 11.4).

Figure 11.4: An Interpretation of a MacArthur Stick with Random Throws

3.   One then "breaks" the stick at the points of the random throws. Thus, making a "broken stick" (Figure 11.5).

Figure 11.5: An Interpretation of a MacArthur Stick Broken

Since this is all rather mysterious, we should pause to note that this procedure of randomly segmenting the stick is done to avoid importing auxiliary assumptions about what may produce changes in a target community's structure into the B-stick model being generated. In other words, the process of generating B-stick models is designed to produce models that permit one to test whether conceiving of niches as non-overlapping and continuous produces useful simulations, and therefore leave out assumptions about which other factors may also influence community structure. The remaining steps in the process of simulating a community with a B-stick model are as follows.

4.   One uses the segments to represent the percentage of available shared resources used by each population being simulated. This is done by assuming that the amount of the total available

resource(s) being used by each population is proportional to each population's relative abundance (Figure 11.6).

Figure 11.6: An Interpretation of a MacArthur Broken Stick with Generated Values

5. Last, one represents the relative abundances of the modelled populations' by arranging the "weighted" line segments on a graph, in order of their relative masses (Figure 11.7).

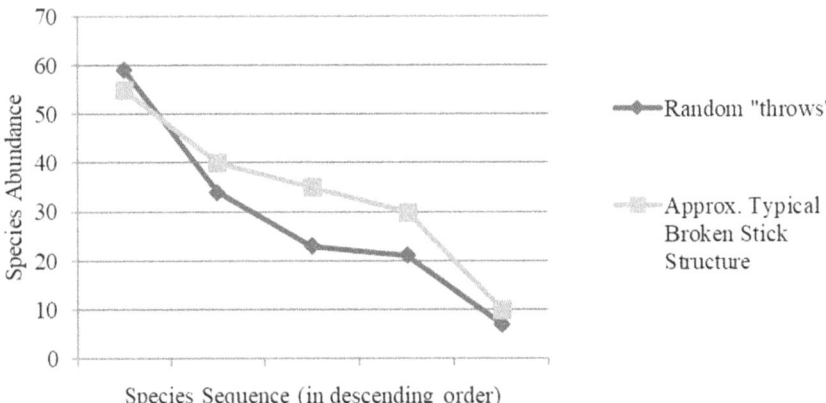

Figure 11.7: A Graphical Interpretation of our Broken Stick Model. From Donhauser (2015); the 'Approx. Typical Broken Stick Structure' approximates to the curves that typically occur when generating B-stick models

By following these steps, one can develop a B-stick model of a target community or possible community given only the number of discrete species comprising it and the number of resources they share. Of course, one cannot simply assume that B-stick models accurately predict actual community abundance schemas, since there is no *prima facie* reason to expect they would. As mentioned, MacArthur (1957) purported to demonstrate that the the hypothetico-deductive method (HD method) is a useful approach to developing predictive ecological principles by showing that B-stick models successfully predict the abundance schemas of some actual communities. He also showed that B-stick models, and the principles they assume, predict actual abundance schemas better than models based on alternative principles.

He did both of these things by comparing B-stick models and other theoretical models with empirical models generated from existing census data on bird populations at particular research sites. In essence, he did this comparison by generating B-stick models and alternative TEMs, and then superimposing the curves produced onto curves generated by graphically plotting existing census data on particular bird communities (MacArthur, 1957, p. 294). Thusly, MacArthur demonstrated that B-stick models match abundance schema curves derived from census data much better than models that assume that niches are overlapping and better than those that assume that they are not continuous (MacArthur, 1957, p. 295).

Since we are looking at MacArthur's project to clarify what is being tested with theoretical models in ecological studies, we will maintain a neutral stance on whether B-stick models are generally useful for predicting the abundance schemas of natural ecological communities. Accordingly, we will not weigh in on whether MacArthur's shows that B-stick models and the principle they assume are in fact predictively powerful, since whether MacArthur actually established this is irrelevant to the question at hand. This is the question of whether MacArthur meant to test more than just the predictive power of B-stick models and the principle they are based on via the HD method. According to the traditional interpretation of the HD method, one is to believe that MacArthur tried to show that communities are in fact comprised of populations that inhabit non-overlapping continuous niches and that B-stick models are representationally accurate characterizations of natural communities. We think this interpretation is misguided, and can show as much by providing evidence that MacArthur's aimed to establish only that B-stick models and the compound principle they assume are instrumentally useful. We contend, moreover, that modern ecologists in general embrace a broadly instrumentalist epistemology.

Why does the traditional view fail? One simple reason why one should not believe that MacArthur (1957) intended to show that B-Stick models and their underlying principle are accepted is that he does not say that this is what he is doing. As is typical in TEM-based research studies, MacArthur never claims to demonstrate that the principles or models he is testing are true or false; which he would probably do if that were his aim. Without looking at what he actually does in the project, one might mistakenly take his claim to be introducing "a more fruitful approach" to ecological research to mean that it is a better way to show that select principles or models should be accepted. However, upon careful reading, it is clear that MacArthur purports only to test how well he can predict abundance schemas with models that assume either that 'niches are overlapping', that they are 'non-continuous', or that they are 'non-overlapping and continuous'. It is also the case that he only explicitly claims to demonstrate that models using B-stick models can produce predictions that closely match data on population abundances. Although he describes what he is doing as testing three different "hypotheses", which might also suggest that he is making substantive claims about the nature of natural communities, there is no indication that these "hypotheses" are supposed to be literally descriptive. In fact, MacArthur does not suggest anywhere that he believes that the winning "hypothesis" is true, corresponds to natural laws, or anything of the sort.

Another reason that one should not believe that MacArthur (1957) intended to show that B-Stick models and their underlying principles are acceptable is that, by design, the HD-testing method he employs does not show that the characterizations being tested are true or false. In fact, ecologists have shown that predictive success "is not evidence that the basic stipulation of nonoverlapping and contiguous niches is correct" by predicting actual abundance schemas with models based on alternative principles as accurately as one can with B-stick models (King, 1964, p. 726). As a more general point, examples of theoretical principles and models that are not descriptive of anything in nature but are nevertheless predictively powerful abound. In light of these things, one must conclude that substantiating models and their underlying principles by showing they accurately predict certain things about certain sorts of natural phenomena simply does not show they are true. Rather, it merely shows that they are, at least sometimes, useful for making predictions about certain sorts of natural phenomena.

The fact that many theoretical models and their underlying principles are noticeably false is good evidence that ecologists are not generally attempting to show that such characterizations accurately describe ecological reality via HD-testing (see Odenbaugh, 2010, p. 159 and 2011, p. 1184). B-stick models and the principles they assume are a case in point, since they are false and fail to accurately characterize most ecological communities in at least three ways. First, because they represent niches as 'non-overlapping resource areas', B-stick models misrepresent how niches and populations can, and usually do, spatially overlap. If interpreting the claim that 'niches are non-overlapping' mereotopologically, one can see that many counterexamples exist (see Smith & Varzi, 1999, pp. 228-230). For instance, grazing animals, like rabbits and deer, that are eating the same grass next to each other, occupy niches that overlap in space. If the claim is understood functionally, in terms of populations playing non-overlapping consumer roles in a community, similar counterexamples abound, since competing populations within a community "function" so as to compete for *the same* resources (such that they are functioning so as to compete for those resources in spatially overlapping areas). Second, B-stick models misrepresent most actual niche topographies by assuming that niches are continuous, since "natural environments are [typically] patchy and the resources within them are heterogeneously distributed" (Codling & Dumbrell, 2012, p. 145). Hence, if interpreting the claim that 'niches are continuous' mereotopologically, there likely exist no (or at least very few) communities that could be comprised of populations that inhabit continuous niches. Interpreting this claim functionally, in terms of consumer roles, is no better. Indeed, since neither populations nor the organisms comprising them can literally consume resources *continuously*, it is hard to make heads or tails of how a 'continuous consumer role' could be instantiated in nature. Third, because B-stick models assume that the amount of resources used by each population is proportional to the relative mass of each modeled population regardless of what species they are, very few natural communities can be literally described by a B-stick model. This is because typical rates of resource consumption vary widely across species (Sugihara, 1980; Tokeshi, 1996). These three points might suggest that MacArthur was mistaken about how natural communities work. However, it's hard to believe that MacArthur – the ecologist who many laud as having made more significant contributions to systems biology than anyone – would have overlooked all of these very obvious facts. Hence, we submit that the fact that his characterizations of communities are false further suggests that he was not trying to show that they are accurate characterizations of natural communities.

A final note that speaks against the traditional interpretation of theoretical ecological research and favors an instrumentalist interpretation is that the processes ecologists use to generate models often include steps designed to purposely produce models that misdescribe ecological phenomena (often to produce simpler and more predictively powerful TEMs). MacArthur's process of generating B-stick models illustrates what is meant by this. Recall that that process includes steps to ensure that factors that are causally operative in producing the relative species abundances within natural communities are *not* taken into account in simulations with B-stick models. The most obvious step taken to ignore such causal factors is the procedure of randomly segmenting the stick so as to avoid importing into a model assumptions about what may produce changes in a target community's structure. By including this step, B-stick models ignore the influences that species composition and diversity play in structuring natural communities, even though these factors significantly impact the way that communities are structured since species-typical consumption rates vary widely. Since it is hard to believe that MacArthur would have overlooked the fact that variability in consumption rates can significantly impact actual abundance schemas, it appears that he introduced steps to ignore this factor to ensure that he tested only whether simulating communities *as if* they are comprised of populations that inhabit non-overlapping continuous niches produces good predictions (MacArthur, 1957, p. 295).

Others have argued that the use, and successful application, of false models in ecology has profound implications for historical debates in the philosophy of science. For example, Jay Odenbaugh (2011) effectively argues that it presents a foil for received forms of scientific realism – according to which we are justified in believing that scientific theories are literally true. On this point, we will here just say that we think the use of false models in ecology, and throughout science, clearly shows that a more sophisticated form of realism than are often endorsed in the philosophical literature are needed.[7]

5. Suggested Modifications for Scientonomy

The extant scientonomic literature does not provide much, if any, attention to models – let alone models used in ecology. Clearly, the models described function much differently than theories that make straightforward claims about reality. However, the precise ways in which current scientonomic theory fails to accommodate these models require acute attention and will be the focus of this section.

As argued, none of the models proposed were meant to be accepted. They do not, and most likely cannot, provide accurate descriptions of anything. Still, we have seen that their motivations stem from their *predictive* powers. In more laymen terms, they are fit to be *used*. According to the currently accepted definition of *use*, provided by Barseghyan (2015, p. 31), being used amounts to being taken as an adequate tool for practical application (Figure 11.8).[8]

Theory Use ≡
A theory is said to be used if it is taken as an adequate tool for practical application.

Figure 11.8: The definition of 'theory use'

This is illustrated with a number of examples including bridge building, winning elections, crop yield increases, and constructing microchips. Strictly speaking, theories aren't physically used in any of these activities; they are 'used' in the sense that their predictions can be helpful in directing the physical acts in constructing artifacts. They are 'used' in a more abstract sense than a hammer or a hand. A great deal hinges on the vagueness of the term 'practical'. The first dictionary definition of 'practical' is that something is practical if it is "concerned

---

[7] One of us is of the view that Anjan Chakravartty's (2007) brand of "semirealism" is the most promising formulation of realism.

[8] See also https://scientowiki.com/Theory_Use_(Barseghyan-2015)

with the actual doing or use of something rather than with theory and ideas". This *contrasts* the usefulness of a hammer, which can be 'actually' (i.e., materially) used rather than a theory. Of course, we need not be restricted by this dictionary definition though this highlights that there is something importantly different about the way theories are used and the way hammers are used. Indeed, if one is a pragmatist or a traditional Marxist, then *all* activity, including theoretical activities like answering questions are, at root, concern use (even if covertly and unconsciously). For example, John Dewey writes that "Science is converted into knowledge in its honorable and emphatic sense *only* in application. Otherwise it is truncated, blind, distorted" (Dewey, 1954, p. 174). Though Dewey's view comes with his own method, it simply reinforces the need to specify what counts as a 'practical' versus 'not practical' activity. We suggest that it would be fruitful to distinguish between two kinds of use. One is *practical theory use* and the other is *epistemic theory use*. *Practical theory use* can be seen as a refined definition of *theory use* in the traditional scientonomic sense. However, it is slightly more specific (Figure 11.9).

Practical Theory Use ≡
A theory is said to be practically useful if it is deemed to be helpful in guidaing practical activities.

Figure 11.9: The definition of 'practical theory use'

A few words are needed here. First, it is clearer than 'adequate tool'. As mentioned, a theory is a 'tool' in a different sense as a hammer and so this phrase may be misleading. Second, it makes explicit reference to *guiding* practical activities.[9] This specifies the sense in which theories are helpful in practical activities. Finally, it avoids the metaphysical ambiguity in the term 'application'. How can a theory, which is abstract, become 'applied' which makes it concrete? If it isn't the *theory itself* that is concrete, then the theory can't be used in the material sense. This is a very difficult question, and one we have no answer for here. Luckily, though, this new definition of practical use dodges these worries.

The models discussed in the previous sections are certainly useful for particular aims such as testing ranges of possibility regarding biological populations in different situations. These uses are *epistemic*; they serve epistemic goals. These goals do not guide practical activities, though they certainly can and are often motivated by practical considerations. To put this point differently, the 'uses' of the models depend on the context. In theoretical ecology, the models are useful

---

[9] A similar formulation is provided by Popper (1970, pp. 260-261).

for providing predictions in a simple and accurate manner. *This* is the epistemic stance scientists take towards these models. This is seen most acutely in our description of Hutchinson – his argument for broken stick models is not that they tell us anything about nature, that is they can be answers to questions about what the world is like, but their predictive accuracy. Of course, these models can *also* be used for practical guidance by, for example, policy makers of various stripes. This practical use is a different context and the models are adjusted and evaluated on slightly different grounds. This point can be made clearly using Barseghyan's toy example of bridge building. General relativity is useful for many purposes, including predictions and explanations, but not useful for bridge building simply because it is too complicated. The stance of whether or not it is 'useful' depends on context. Of course, there may be many contexts that could be specified; but we will begin this task by simply distinguishing between practical theory use, as defined above, and *epistemic theory use* (Figure 11.10).

Epistemic Theory Use ≡
A theory is said to be epistemically useful if it is deemed suitable for a set of epistemic goals.

Figure 11.10: The definition of 'epistemic theory use'

Sometimes, it is the case that the fulfilling of an epistemic goal is a part of a method for accepting the theory. For example, fulfilling the goal of novel predictive accuracy is sufficient for being accepted by a community who holds the HD method. However, there is no evidence that the communities in theoretical ecology at the time employed anything like HD method and such a reconstruction seems highly unlikely given what we have already argued in the previous sections. Still, predictive accuracy was held to be a virtue of these models. Thus the fulfilling of this epistemic goal made them deemed to be epistemically useful, but not necessarily acceptable (cf. Jacquart, 2016).

The second suggestion comes by way of figuring out how to introduce the notion of a model into the scientonomic taxonomy or whether it is reducible to the scientonomic notion of a 'theory'. According to the current definition of *theory*, suggested by (Sebastien, 2016), a theory is a set of propositions (Figure 11.11).[10]

---

[10] See also https://www.scientowiki.com/Theory_(Sebastien-2016)

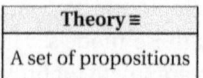

Figure 11.11: The definition of 'theory'

Of course, the literature on the relationship(s) between models and theories understood as sets of propositions is enormous and impossible to cover here. However, to understand this tricky point here, we must clarify a related issue. Namely, what would it mean for a proposition to be 'reduced' to a model?

L-V models obviously rely on general propositions such as "predators always kill prey at rate x" that are not true of any particular populations (see Donhauser, 2017). But this does not mean that the L-V model *is* a set of propositions – it simply means that assumptions they rely on are expressible as propositions. Can such a model, or any model, be *logically* or *mathematically* reduced to a set of propositions? The overwhelming response to these questions is a resounding 'no!' for numerous reasons that are too complicated to rehearse here (see, e.g., Suarez, 2015 and Knuuttila, 2011). In brief, models are particular artifacts of particular modelling practices that represent actual or possible phenomena that are being considered by researchers partially and approximately, and which often rely on all sorts of distortions, fictions, and context-sensitive interpretive elements that help those constructing the models understand select aspects of target phenomena. So, because models rely on complex, and historically contingent, relations between researchers, sets of propositions, and representational artifacts of different sorts, it is impossible in practice to reduce them to sets of propositions. Yet, again, any L-V model can be *expressed* propositionally – just as we have expressed exemplars of such models in discussing them throughout this chapter.

Finally, it is worth pointing out the advance made in the scientonomic definition of *theory acceptance*. The initial definition of the term was proposed by Barseghyan (2015, p. 39) and stated that theory acceptance amounts to being considered the best available description of the respective object (Figure 11.12).[11]

Theory Acceptance ≡
A theory is said to be accepted if it is taken as the best available description of its object.

Figure 11.12: Barseghyan's original definition of 'theory acceptance'

---

[11] See https://www.scientowiki.com/Theory_Acceptance_(Barseghyan-2015)

The definition was then refined by Sebastien (2016) to ensure that not only descriptive but also normative theories can be accepted (Figure 11.13).[12]

Theory Acceptance ≡
A theory is said to be accepted if it is taken as the best available description or prescription of its object.

Figure 11.13: Sebastien's definition of 'theory acceptance'

Finally, according to the currently accepted definition of the term (Barseghyan, 2018, p. 31), a theory is said to be accepted by an epistemic agent, if that agent considers the theory to be the best extant answer to its respective question (Figure 11.14). The main motivation here was to provide a definition broad enough that would be applicable to theories of *all* types.[13]

Theory Acceptance ≡
A theory is said to be accepted by an epistemic agent if it is taken as the best available answer to its respective question.

Figure 11.14: Barseghyan's new definition of 'theory acceptance'

While we will not defend this definition here, we should remark that our case studies provide further support for this change. The models described in this paper do not describe anything. Indeed, they are *intentionally false* or, in other cases, only *partly true*. This is a common feature of models and their use of idealizing and abstracting assumptions. They don't even try to be descriptive, though their use of abstractions and idealizations make them suitable for various cognitive ends. But models, like anything really, can be construed as answers to questions and evaluated as better or worse relative to the aims they attempt to fulfill.

However, a problem remains. Currently, in scientonomy, theories can either be descriptive, normative, or a definition. This, at the very least, oversimplifies the picture. Abstract propositions such as "Rabbits are thermal vectors" are *partly* true – they may be said to be descriptive of a single aspect of rabbits. They may be reconstrued as definitions that are necessary for the functionality of the model, but this doesn't seem right either since they are representational. Abstract statements like this are abound in science, but they are not universal; not all propositions are abstractions in the same sense. Peter Godfrey-Smith (2009)

---

[12] See https://www.scientowiki.com/Theory_Acceptance_(Sebastien-2016)
[13] See https://www.scientowiki.com/Theory_Acceptance_(Barseghyan-2018)

helps make this point clear. Some abstractions leave out details, some *imagine those details don't exist.* For example, Newton's laws abstract away from real-world phenomena like friction. They assume that friction doesn't exist; a counterfactual assumption. But still, Newton's laws describe a *part* of mechanical systems, and a *full* description is given when combined (and sometimes altered) along with other variables.[14] Thus we propose the notion of *abstract theory* as a subtype of descriptive theory (Figure 11.15).

Abstract Theory ≡
A descriptive theory that purports to describe one aspect of its object while presupposing other theories that are false.

Figure 11.15: The definition of 'abstract theory'

For idealizing propositions, such as "populations occupy non-overlapping niches" or "ecological networks *organize* to maximize exergy", they could be said to be descriptive theories but *false.* This would lead to a confusing state of affairs to explain scientists stances towards idealizations. Could scientists accept these propositions? Well, like all propositions, it can be converted into an answer to a question. To evaluate an idealization as "the best" via a method is a bit trickier. Recall that currently *method* is defined as a set of criteria for theory evaluation (Barseghyan, 2018, p. 32).[15]

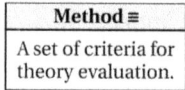

Method ≡
A set of criteria for theory evaluation.

Figure 11.16: The definition of 'method'

Many methods, including those listed as examples, aim at something like "truth" or "empirical adequacy". But idealizations do not have the same aim; they aim at being epistemically useful. So methods do not necessitate theory acceptance, but a theory's epistemic usefulness. We don't think this necessitates any changes in the taxonomy, but it underlines that the category of epistemic use is epistemically useful (*rimshot*) for describing scientists' attitudes towards idealizations.

---

[14] This assumes, against Nancy Cartwright (1994), that laws are literally false.
[15] See https://www.scientowiki.com/Method_(Barseghyan-2018)

6. Concluding Remarks

We hope that our suggested modifications, if accepted, will lead to a more fruitful and refined ontology of scientific change. Before finishing, it is worth pointing out that a hole has been uncovered in this paper though we did not elaborate on it or suggest solutions. Ontologically speaking, what is a theory? Similarly, what is a model? To be clear on what a theory is and how agents may interact with it is necessary for a well-defined ontology. Is a theory something 'in the head' as Quine famously argued? Is a theory floating about in Fregean heaven? Answers to questions such as these are necessary if we are to have a precise sense of what it means to 'use' a theory and similar implications may follow for other epistemic stances. After all, scientonomy is the "science of science" which implies that its objects are all parts of the natural world. This difficult topic, however, we will leave for further research.

Bibliography

Barberousse, A., Morange, M., & Pradeu, T. (Eds.) (2009). *Mapping the Future of Biology: Evolving Concepts and Theories.* Springer.

Barseghyan, H. (2015). *The Laws of Scientific Change.* Springer.

Barseghyan, H. (2018). Redrafting the Ontology of Scientific Change. *Scientonomy*, 2, 13-38.

Bender, E. A., Case, T. J., & Gilpin, M. E. (1984). Perturbation Experiments in Community Ecology: Theory and Practice. *Ecology*, 65(1), 1-13.

Cartwright, N. (1994). Fundamentalism vs. the Patchwork of Laws. *Proceedings of the Aristotelian Society*, 94, 279-292.

Chakravartty, A. (2007). *A Metaphysics for Scientific Realism.* Cambridge University Press.

Codling, E. & Dumbrell, A. (2012). Mathematical and Theoretical Ecology: Linking Models with Ecological Processes. *Interface Focus*, 2(2), 144-149.

Coffaro, G., Bocci, M., & Bendoricchio, G. (1997). Application of Structural Dynamic Approach to Estimate Space Variability of Primary Producers in Shallow Marine Water. *Ecological Modelling*, 102(1), 97-114.

Cohen, J. & Newman, C. (1985). A Stochastic Theory of Community Food Webs: I. Models and Aggregated Data. *Proceedings of the Royal society of London. Series B. Biological sciences*, 224(1237), 421-448.

Dewey, J. (1954). *The Public and its Problems.* Ohio University Press.

Donhauser, J. (2014). On How Theoretical Analyses in Ecology Can Enable Environmental Problem-Solving. *Ethics & the Environment*, 19(2), 91-116.

Donhauser, J. (2015). *A Philosophy of Theoretical Ecology for Environmental Policy.* Doctoral Dissertation, University at Buffalo.

Donhauser, J. (2016). Theoretical Ecology as Etiological from the Start. *Studies in History and Philosophy of Science Part C: Studies in History and Philosophy of Biological and Biomedical Sciences*, 60, 67-76.

Donhauser, J. (2017). Differentiating and Defusing Theoretical Ecology's Criticisms: A Rejoinder to Sagoff's Reply to Donhauser (2016). *Studies in History and Philosophy of Science Part C: Studies in History and Philosophy of Biological and Biomedical Sciences*, 63, 70-79.

Donhauser, J. & Shaw, J. (2019). Knowledge Transfer in Theoretical Ecology: Implications for Incommensurability, Voluntarism, and Pluralism. *Studies in History and Philosophy of Science Part A*, 77, 11-20.

Godfrey-Smith, P. (2009). Abstractions, Idealizations, and Evolutionary Biology. In Barberousse, Morange, & Pradeu (Eds.) (2009), 47-56.

Holt, R. & Pickering, J. (1985). Infectious Disease and Species Coexistence: A Model of Lotka–Volterra Form. *The American Naturalist*, 126(2), 196-211.

Hutchinson, G. (1948). Circular Causal Systems in Ecology. *New York Academy Sciences Annals*, 50, 221-246.

Jacquart, M. (2016). *Similarity, Adequacy, and Purpose: Understanding the Success of Scientific Models*. Doctoral Dissertation, University of Western Ontario.

Jørgensen, S. (2000). The Tentative Fourth Law of Thermodynamics. In Jørgensen & Müller (Eds.) (2000), 161-175.

Jørgensen, S. & Müller, F. (Eds.) (2000). *Handbook of Ecosystem Theories and Management*. CRC Press.

Jost, C., Devulder, G., Vucetich, J., Peterson, R., & Arditi, R. (2005). The Wolves of Isle Royale Display Scale-Invariant Satiation and Ratio-Dependent Predation on Moose. *Journal of Animal Ecology*, 74(5), 809-816.

Keller, E-F. (2008). Organisms, Machines, and Thunderstorms: A History of Self-Organization, Part One. *Historical Studies in the Natural Sciences*, 38(1), 45-75.

Knuuttila, T. (2011). Modelling and Representing: An Artefactual Approach to Model-Based Representation. *Studies in History and Philosophy of Science*, 42, 262-271.

Lotka, A. (1925). *Elements of Physical Biology*. Williams & Wilkins Company.

Mäkelä, A. & Hari, P. (1984). Interrelationships Between the Lotka-Volterra Model and Plant Eco-Physiology. *Theoretical Population Biology*, 25(2), 194-209.

MacArthur, R. (1957). On the Relative Abundance of Bird Species. *Proceedings of the National Academy of Sciences of the United States of America*, 43(3), 293-295.

Magnus, P. D. & Busch, J. (Eds.) (2010). *New in Philosophy of Science*. Palgrave Macmillan.

Mitsch, W. & Jørgensen, S. (2004). *Ecological Engineering and Ecosystem Restoration*. Wiley.

Müller, F. & Leupelt, M. (1998). *Eco Targets, Goal Functions, and Orientors*. Springer.

Odenbaugh, J. (2010). Philosophy of the Environmental Sciences. In Magnus & Busch (Eds) (2010), 155-171.

Odenbaugh, J. (2011). True Lies: Realism, Robustness, and Models. *Philosophy of Science*, 78(5), 1177-1188.

Odenbaugh, J. & Alexandrova, A. (2011). Buyer Beware: Robustness Analyses in Economics and Biology. *Biology & Philosophy*, 26(5), 757-771.

Peters, R. (1991). *A Critique for Ecology*. Cambridge University Press.

Popper, K. (1970). Reason or Revolution? *European Journal of Sociology/ Archives Européennes de Sociologie*, 11(2), 252-262.

Rant, Z. (1956). Exergy, A New Word for Technical Available Work. *Forschungen im Ingenieurwesen*, 22(1), 36-37.

Reiners, W. & Lockwood, J. (2010). *Philosophical Foundations for the Practices of Ecology*. Cambridge University Press.

Sebastien, Z. (2016). The Status of Normative Propositions in the Theory of Scientific Change. *Scientonomy*, 1, 1-9.

Slack, N. (2011). *G. Evelyn Hutchinson and the Invention of Modern Ecology*. Yale University Press.

Smith, B. & Varzi, A. (1999). The Niche. *Noûs*, 33(2), 214-238.

Suárez, M. (2015). Deflationary Representation, Inference, and Practice. *Studies in History and Philosophy of Science Part A*, 49, 36-47.

Sugihara, G. (1980). Minimal Community Structure: An Explanation of Species Abundance Patterns. *American Naturalist*, 116(6), 770-787.

Susani, L., Pulselli, F., Jørgensen, S., & Bastianoni, S. (2006). Comparison Between Technological and Ecological Exergy. *Ecological Modelling*, 193(3-4), 447-456.

Tett, P. & Wilson, H. (2000). From Biogeochemical to Ecological Models of Marine Microplankton. *Journal of Marine Systems*, 25(3-4), pp.431-446.

Thornley, J. H. M., Bergelson, J., & Parsons, A. J. (1995). Complex Dynamics in a Carbon-Nitrogen Model of a Grass-Legume Pasture. *Annals of Botany*, 75(1), 79-94.

Tokeshi, M. (1996). Power Fraction: A New Explanation of Relative Abundance Patterns in Species-Rich Assemblages. *Oikos*, 75(3), 543-550.

Turchin, P. (2001). Does Population Ecology Have General Laws? *Oikos*, 94(1), 17-26.

van Fraassen, B. (2008). *The Empirical Stance*. Yale University Press.

Wiener, N. (1948). *Cybernetics or Control and Communication in the Animal and the Machine*. MIT Press.

Wilson, B. (1993). Would We Recognise a Broken-Stick Community if We Found One? *Oikos*, 67(1), 181-183.

Zhang, J., Gurkan, Z., & Jørgensen, S. (2010). Application of Eco-Exergy for Assessment of Ecosystem Health and Development of Structurally Dynamic Models. *Ecological Modelling*, 221(4), 693-702.

Chapter 12

The Relative A Priori as a Model of Radical Conceptual Change in Science

David J. Stump

University of San Francisco

Abstract: Relative, pragmatic, or dynamic theories of the a priori have been considered by many philosophers of science. I present these theories as a model of how radical conceptual change occurs during a scientific revolution. When elements of a theory that are considered to be a priori or constitutive change, we have a revolutionary change that requires rethinking all of a scientific practice. Given that conceptual change is the flashpoint for discussion of the issues of incommensurability, the rationality of scientific change and relativism, by exploring theories of the a priori I show how radical conceptual change can occur and defend the rationality of scientific change. The viewpoint adopted avoids commitment to traditional a priori knowledge and to metaphysics, while still acknowledging that there is an important element in science that cannot simply be described as empirical. I present evidence to show that the model of scientific change can be applied widely.

Keywords: conceptual change; relative a priori; pragmatism

***

1. A Model of Conceptual Change in Science

In science, there are theories and principles that are taken for granted before empirical inquiry can begin. While these theories and principles may have been confirmed empirically, some fundamental principles or laws and all of the mathematics upon which science depends have a more problematic basis, since these principles are very difficult to conceive of as being empirically grounded. Thus, some of the principles, and all of the mathematics, appear to be examples of a priori knowledge. They serve as constitutive elements in science that play a special role in scientific theories, given that they are necessary preconditions to further inquiry. I suggest that when the elements of scientific theories that have been considered to be a priori change, there is a radical revolution in science, a conceptual change.

Conceptual change during scientific revolutions is a major topic in the philosophy of science but the concept has proven to be elusive and has also raised the issue of relativism, especially in the wake of Kuhn's *Structure of Scientific Revolutions* (1962). Given that it is the most extreme kind of scientific change, conceptual change is the most likely to lead to claims that science is relative, that different scientific practices are incommensurable, that the development of science is discontinuous rather than cumulative, and that scientific change is not rational (Nersessian, 1989; Arabatzis & Kindi, 2008, p. 350). In short, conceptual change has been the flashpoint for issues around scientific revolutions.[1] By radical conceptual change, I have in mind instances where a scientific change required a reinterpretation or reevaluation of theories and practices. As part of his argument that the development of science is discontinuous, rather than cumulative, Kuhn famously said that scientific revolutions are destructive, as well as constructive (Kuhn, 1962, pp. 7, 66). A practice or theory that existed before a scientific revolution may have to be given up, such as astronomy based on an earth-centered model of the solar system or chemical theories and practices based on phlogiston.[2] Something that was explained by a former paradigm may be left unexplained in a new paradigm, at least temporarily. For example, Scholastic theories explained continued motion in a straight line, whereas Newton's theory took such motion as natural and not in need of explanation. The overthrow of Scholastic theories of motion by Newton was seen as instructive since understanding Newton's laws required a new way of thinking, not merely the acknowledgment of a new fact. Even in the twentieth century, after Newton's laws were replaced with those found in the special and general theories of relativity, a lively debate over the status of Newton's laws of motion took place in the philosophical literature, where the introduction of the first law was seen as an extremely important example of conceptual change (Hanson, 1958, 1965; Toulmin, 1961; Nersessian, 1989). Looking at the constitutive elements in science gives us a way of understanding deep conceptual change in science, which again seems to be neither simply empirical nor unproblematically rationalistic. Constitutive elements in science

---

[1] See Sankey, 2019, for a current survey of the relativistic aspects of conceptual change as presented by Kuhn and Feyerabend. Also see Nersessian, 2003, and for a contemporary account of the development of scientific concepts, Nersessian, 2008.

[2] This has been called "Kuhn loss" in the literature. See Shaw, 2020, for an overview and evaluation of the significance of the phenomenon.

seem to be isolated from ordinary methods of justification, given that they have to be in place before an inquiry can begin.

My project of understanding conceptual change in science through an examination of the constitutive elements of scientific practice is compatible with scientonomy, the search for laws of scientific change, although my focus is narrower, given that I look only at a way of understanding radical conceptual change, rather than all change. However, this focus is justified given that radical conceptual change is the most contentious kind of scientific change, stemming from Kuhn's seminal treatment of it in *Structure* (1962). I accept all the arguments given in *The Laws of Scientific Change* that a theory of scientific change is possible (Barseghyan, 2015, ch. 2). The extent to which a theory of scientific change has been established, that is, how well the theory holds up to detailed historical studies, is an empirical question that I will not discuss, presenting instead a way of understanding what we mean by radical conceptual change in science. My claim is that when constitutive elements of a theory or practice change, the theory or practice needs to be rethought and reconstructed from the ground up.

Broad philosophical models of any aspect of science may be questioned on the grounds that they run roughshod over the history of science, especially the contingent, particular context of a scientific practice so important to understanding any episode in the history of science. Some might argue that we do not need philosophical models of science at all, espousing an anti-theory view in which all history is local and ruling out the possibility of generalizing claims about science and its history. However, the idea that history can be free of philosophical presuppositions is naïve, just as is the view that science can be value-free. An approach of some kind is necessarily adopted when one studies science, historically or otherwise. I introduce one family of philosophical models of conceptual change in science, referred to here as theories of the constitutive elements in science, that reflects a more accurate image of science than that of competing philosophical models such as Quinean empiricism, while claiming that these models can be used without violating the standards of the historiography of science and further argue that theories of the constitutive elements in science can be formulated without falling into relativism, idealism, or social construction.

2. The Relativized, Dynamic, Pragmatic, or Functional A Priori

The idea of changing a priori starts with Reichenbach's briefly held view in *The Theory of Relativity and A Priori Knowledge* (1965). Reichenbach notes that there are two elements to the Kantian a priori. A priori statements are necessary, but also constitutive. Reichenbach suggests giving up on necessity, but keeping the idea of constitution, thus advocating for an a priori that can change. Paolo Parrini (1983, 1998) and Michael Friedman (1999, 2001) each revived Reichenbach's idea of a dynamic a priori, defending the idea that conceptual revolutions occur in science when there is a change in what had been taken to be a priori knowledge. We find a parallel development in pragmatism, centered on the work of C. I. Lewis (1923), but existing as well in the work of Peirce and Dewey. Arthur Pap developed the pragmatic conception of the a priori in his dissertation, calling it a functional theory of the a priori (Pap, 1946). Similar views were developed independently by Ernst Cassirer (1923), Thomas Kuhn (2000, p. 264), Ian Hacking (1992), and briefly, Hilary Putnam (Tsou, 2010). All theories of the constitutive elements in science distinguish between empirical laws and constitutive principles, but as Pap emphasized, what is empirical and what is constitutive can change over time or in different contexts. When constitutive elements change, we have a conceptual revolution since the aims, standards of explanation, and problems being addressed can all change.

The example that Friedman has worked through in great detail is the history of mechanics and gravitational theory from Newton to Einstein (Friedman, 2001). As Friedman shows, Newton's laws (or their relativistic counterparts) and the mathematics used in physics are the constitutive elements, while Newton's law of gravity and Einstein's field equations are empirical. The constitutive elements, of course, changed from Newtonian theory to Einstein's, in which space is no longer Euclidean but, rather, has variable curvature and Newton's second law receives a relativistic "correction" so that it is consistent with the principle that nothing can travel faster than the speed of light. As Friedman puts it (Friedman, 2008, p. 251):

> Thus, for example, whereas Euclidean geometry and the Newtonian laws of motion were indeed necessary presuppositions for the empirical meaning and application of the Newtonian theory of universal gravitation (and they were therefore constitutively a priori in this context), the radically new mathematical and physical framework consisting of the Riemannian theory of manifolds and the principle of equivalence defines an analogous system of necessary presuppositions in general relativity.

Some constitutive principles must necessarily be in place, although they can change rather dramatically from one scientific theory to another.

3. Constitutive Elements in Science

Constitutive elements were often taken to be a priori. Indeed, one way to organize much of twentieth-century philosophy of science is to read it as a series of debates over what had been considered a priori knowledge. Although synthetic a priori knowledge was officially rejected by the Vienna Circle and their followers, parts of knowledge that had been considered a priori by Kant, such as geometry, space and time, causality, and the basic principles of physics, were widely discussed throughout the twentieth century, and statements about them have often been given a special role, whether as conventions or as the hard core of scientific theories. One would think that Quine's critique of the analytic/synthetic distinction would have put the final nail in the coffin of a priori knowledge and that a holism in which all scientific statements are justified empirically has replaced the notion of any special status for what was formerly considered to be a priori, but in fact, replacement theories of the a priori were developed throughout the twentieth century, as noted above. Even though Kant thought of a priori knowledge as fixed and absolutely certain, these theories of the relative, dynamic, functional, or pragmatic a priori see the a priori as changing during scientific revolutions. Indeed, in Pap's view, that which is constitutive changes from one context to another.

After comparing these different versions of the a priori, I argue for a pragmatic model that emphasizes the constitutive elements in science without reference to an a priori (Stump, 2015). The term 'a priori' is misleading given that what is called the dynamic or pragmatic or functional a priori is not actually a priori at all in the traditional sense. Instead, we have various theories of the constitutive elements in science, Kant's, in which the constitutive elements actually are a priori in the traditional sense, that is, necessary and fixed, and the others, in which at least some of the constitutive elements are not fixed, so that we can understand the conceptual change in science as changes in these constitutive elements. Divorcing the issue of conceptual change from a priori knowledge brings considerable clarity and moves the discussion beyond the issues of empiricism and rationalism and toward an analysis of science as a practice. While still emphasizing the importance of the constitutive element in science, I stay closer to naturalism than to the neo-Kantian position advocated by Friedman, holding a pragmatic view that the constitutive elements are principles and theories that are necessary preconditions for the possibility of a science.

A pragmatic understanding of the constitutive elements in science should be seen as a deflationary strategy that leaves a core of beliefs about constitutive elements in place, but refuses to see them as transcendental or otherwise outside of the realm of science. Regarding constitutive elements of science, we can agree on the need to begin our inquiry from some principles, but we do not need to claim that they are certain or that they are known by some special intuition, or even that they inhabit a philosophical arena that is separated from science. Constitutive elements can function as what was once called a priori knowledge, even if they are no longer a priori in the traditional sense. On the other hand, we cannot claim that these constitutive elements are just like any other part of empirical inquiry, given that they make science possible, playing a special role as an ineliminable part of an empirical physical theory, even if they are not directly testable.

While it is possible to defend Carnap or even Kant from Quinean holistic empiricism, I take a different approach, arguing that the former a priori should be treated as constitutive, by which I mean that some elements of physical theory have a unique epistemological status in that they are necessary in order to begin empirical inquiry. One must adopt these elements of science before empirical inquiry, so they, therefore, function as a priori parts of our physical theory, chosen for conceptual or pragmatic reasons prior to any empirical testing. Although I am willing to concede to Quine that most of these constitutive elements of science are ultimately empirical, the picture of knowledge that I develop is very different from that developed in Quinean holism in that categories of knowledge can be differentiated. While he admits that some elements of empirical theory are much less likely to be revised than others, differentiating between what he calls the hard core and the periphery, Quine underestimates the asymmetric relation between the constitutive elements and the rest of science. It is not simply that the periphery is more likely to be revised than the hard core but, rather, that the statements of the periphery cannot even be stated, let alone tested, without the constitutive elements functioning as a necessary precondition.

Although discussions of constitutive elements in science (and of the relative, dynamic, functional, or pragmatic a priori) have followed Reichenbach and Friedman in being centered on mathematical physics, specifically theories of space and time, recent work shows that there are constitutive elements in other areas of science. Michele Luchetti shows how the Hardy-Weinstein law plays a constitutive role in evolutionary biology, arguing that we can learn lessons about its function that paint a slightly different picture of constitutive elements relative to that given by Friedman (Luchetti, 2021). Catherine Herfeld shows how understanding the rationality principle in economics as a constitutive principle clarifies its role and explains the long-standing controversy over the status of the principle, given that it seems to be neither empirical nor a priori (Herfeld, 2021). Thus, the model of science as founded on constitutive principles has the potential to be very widely applicable to many areas of science, not just to mathematical physics.

4. Michael Friedman's Rescue of the Rationality of Scientific Revolutions

In order to account for a conceptual change in science, Friedman revived Reichenbach's idea of a changing a priori, using it to show that conceptual revolutions occur in science when there is a change in what had been taken to be a priori knowledge. Like other theories of the constitutive elements in science, Friedman's dynamic theory of the (former) a priori understands conceptual change in science as change in fundamental presuppositions that are required for the practice of science. What is unique about Friedman's account lies in the special role that he assigns to philosophy (Friedman, 2001, p. 66). He sees the constitutive and the empirical elements in scientific theories as layers, adding a third, that of philosophical meta-paradigms or meta-frameworks that provide the intellectual context in which scientists work and that are essential to the development of new constitutive or a priori elements of a theory that develop during a revolution (Friedman, 2008, p. 251):

> Moreover, what makes the latter framework constitutively a priori in this new context is precisely the circumstance that Einstein was only able to arrive at it in the first place by self-consciously situating himself within the earlier tradition of scientific philosophy represented (especially) by Helmholtz and Poincaré – just as this tradition, in turn, had earlier self-consciously situated itself against the background of the original version of transcendental philosophy first articulated by Kant.

What exactly is the problem that Friedman's philosophical layer is supposed to solve? The meta-frameworks are supposed to provide continuity that is lacking in the sciences. Friedman sets out this problem as stemming directly from Kuhn's account of scientific revolutions. No doubt that philosophy plays a big role in certain kinds of scientific disputes, but not the role that Friedman describes. In the first place, science is already continuous enough, so a new philosophical layer is not required to establish the continuity of science through revolutionary change. In the case of relativity, there are transitional figures who are neither purely Newtonian nor purely Einsteinian – figures such as Poincaré and Lorentz. In the second place, I want to resist at all costs the idea that philosophy can be set up as an independent and prior discipline that somehow grounds the sciences. On this point, the naturalists are correct, there is no first philosophy. Friedman is on record opposing naturalism, though mainly the Quinean version (Friedman, 1997). There are many forms of naturalism, but in some versions of the naturalist position, science and philosophy should be seen as continuous, that is, as mutually influencing each other. Friedman should be sympathetic with that, since much of his work has dealt precisely with the interaction between philosophy and science, so it is rather surprising that he gives philosophy a special role in the *Dynamics of Reason*.

No such philosophical meta-frameworks exist, especially not one that meets Friedman's goal of obtaining "an ideal state of maximally comprehensive communicative rationality in which all participants in the ideal community of inquiry agree on a common set of truly universal, trans-historical constitutive principles" (2001, p. 67). As Hasok Chang (2008) points out, there is a question of whether universal agreement should even be a goal or whether some kind of pluralism may be preferable, but my point is that there are no truly universal, transhistorical principles, especially not in philosophy, nor do we need them to be rational (Uebel, 2012). We can make rational choices with rather mundane principles that are located in a particular time and place. We sometimes may be led to an impasse or a situation where we do not have enough evidence to settle a dispute or to answer a question, when we simply have to wait for more evidence to accumulate over time.

Put most strongly, I would argue that there is simply no problem to be addressed by recourse to a transcendental realm of philosophy – scientific revolutions can be rational even without the continuity that this layer of discourse provides. There is no reason to accept Kuhn's view on face value, as Friedman apparently has. A non-foundationalist account of science has the resources to rationally ground theory choice in scientific revolutions. I have in mind views such as those found in the works of Larry Laudan (1977, 1987) and Dudley Shapere (1980, 1984), and I defend them in some detail against the charge of relativism in an early article (Stump, 1991). The same argument applies in the context of my criticism of Friedman's dynamic a priori.

5. Conceptual Change in Science can be Objective without Absolutes

In a recent paper, I argue that denying absolutes does not lead to relativism because "middle ways" exist that have been explored especially by pragmatists (Stump, 2021). My strategy is to look at the elements that define a relativist stance and show where the pragmatist disagrees with the relativist. Defending the pragmatic notion of experience will be central, given that relativists reject the idea that experience can play a role in justifying belief. One of the lessons of pragmatism is that universal and fixed principles are not necessary for objective knowledge, even if we reject all absolutes and start from the premise that everything is historical, contingent, and situated. I will sketch the central argument given in the paper here.

Consider Ian Hacking's styles of reasoning and Martin Kusch's critique of them. Kusch takes up this issue of relativism in section 3.2 of his detailed and insightful critical analysis of Hacking's historical ontology (Kusch, 2010, p. 166). While I agree with many of the points raised by Kusch, he misses the role that experience plays in Hacking's argument, leading to a charge that Hacking is committed to a form of relativism. Like my pragmatic fallibilist, I would say that Hacking's position is that they are relative in one sense and not in another. Of course, it is true that judgments can only be formulated within a style of reasoning, but once that style is adopted, the grounds for judgment are objective. In pragmatist terms, an experiment either works or not, and whether it does is not built into the style of reasoning in advance. Kusch does not take Hacking's insistence that styles of reasoning tell us what is up for grabs as true or false, rather than what is true seriously enough.

> When the question is a live one, and there is a context in which there are ways of addressing the question, or even methods of verification for possible answers, then aspects of the world determine what the answer is, even though only people in a scientific society find out the answer (Hacking, 2000, p. S69).

In empirical science, setting out the basic laws and definitions does not determine what is true. You have to go out and do things – experiment, build things, intervene in nature – in order to find out what is true. Hacking's point is that in the empirical sciences all that the style of reasoning tells you is what is a possible candidate for truth or falsity, not what is true or false. The style gives you the words, the concepts, and even the physical practices that you need in order to express yourself and intervene in the world. In some sense, the style limits us, but in a much stronger sense, it enables us to carry out the practice of science.

Is Hacking then relying on some form of empirical "given" that is an absolute? If so, does it not fall prey to the arguments given by Wilfrid Sellars concerning the Myth of the Given (Sellars, 1956)? The answer to both questions is no, but it will take some argument to show why. Fortunately, Steven Levine just published a book on this issue and Jim Garrison an article, so I refer to them to work out the details rather than rehearse all of the arguments relating to classical and to neo-pragmatism (Levine, 2019; Garrison, 2019). In brief, Levine and Garrison both argue that Rorty and Brandom were wrong to take Sellars' argument as undermining the concept of experience that we get in William James and John Dewey, as well as in Peirce. Rorty claimed that empiricism is a dead end, and that we have no choice but to take the linguistic turn to mean that truth depends on agreement. Most importantly, Levine argues compellingly that we can recover a notion of experience that does not leave us in the linguistic realm and that we can do so without any claim of absolutes. Garrison does so as well, focusing on Dewey.

My own argument stays closer to philosophy of science and, indeed, is a claim about how science works. If we look at current and historical scientific practice, we will not encounter the kind of stalemates that those advocating relativism claim to find in the history of science, especially not in the long term. There can be independent constraints on theory choice even if we accept the view that all observation is theory-laden (Stump, 1991). These constraints can result in objective choice since the theories that are presupposed can be epistemically independent of those under test. Furthermore, the fact that our knowledge is situated in a particular time and place, that we use a particular vocabulary, and that we begin with a set of presuppositions does not imply that our knowledge is relative. The final outcome of a scientific practice is born of an interaction with the material world, not made up whole cloth by scientists. I am not claiming that all scientific practice will converge on a single answer, as a realist might. Rather, I am merely pointing out that scientists are constrained by what they are actually able to construct materially. Nevertheless, we cannot escape the fact that our knowledge is the product of individuals working in a particular context, at a particular time and place, and with the conceptual and material tools that they have inherited from their predecessors. When following historians of science in attempting to understand the views of scientists from other periods on their own terms, we do not need to become relativists, even moderate ones, to take historical views seriously. If we are fallibilists, we accept that there are no universal and fixed foundations for our knowledge and we can therefore understand how there could be a conceptual change in science when the foundations are changed; however, we do need to acknowledge the special role that constitutive elements in science play in order to understand the conceptual changes that can occur during a scientific revolution.

6. A Priori without Metaphysics

The idea that metaphysics is in some sense inescapable has a long history. An old and well-known expression of this idea comes from the British philosopher F. H. Bradley, who remarks that "The man who is ready to prove that metaphysical knowledge is wholly impossible ... is a brother metaphysician with a rival theory of first principles" (1925, p. 1). This quote is discussed by A. J. Ayer in *Language Proof and Logic* (1950, p. 34) where he claims, rather implausibly, that he can avoid the problem by treating metaphysics as linguistically meaningless. In a letter to the publisher Flammarion, Henri Poincaré has a slightly different way of expressing the point that one cannot avoid metaphysics: "Now, those who regard metaphysics as out of style since Auguste Comte will say that there cannot be a modern metaphysics. However, the negation of all metaphysics is still a metaphysics, and it is precisely what I call modern metaphysics".[3] Recently, there have been defenses of metaphysics in the context of philosophy of science, with an organization, the Society for

---

[3] "Maintenant, ceux qui regardent la métaphysique comme démodée depuis Auguste Comte, me diront qu'il ne peut y avoir de métaphysique moderne. Mais la négation de toute métaphysique, c'est encore une métaphysique, et c'est précisément là ce que j'appelle la métaphysique moderne" (Walter et al. (Eds.), n.d.).

the Metaphysics of Science, and conferences, special issues of journals, books, etc. The society itself makes a rather strong disclaimer:

> Though we conceive of the metaphysics of science in a broad manner, it nonetheless contrasts, in both its object phenomena and methodology, with other forms 'metaphysics' has recently taken. In the twentieth century, 'metaphysics' became a dirty word after it was associated with the largely transcendental and a aprioristic approaches of Kantians and Hegelians which were assailed by the Positivists, Wittgensteinians, ordinary language philosophers and others. The metaphysics of science is neither transcendental nor aprioristic since it takes its foundation in the sciences, and it thus also contrasts with much recent analytic metaphysics, though it overlaps with some.[4]

One might think that a viewpoint that held that there is an a priori element in science (or even just a constitutive element) would be committed to some form of metaphysics in science. Given that I defend above various theories of the relative or dynamic a priori, I should clarify in precisely what sense these types of theories are committed to rationalist forms of argument and a priori and/or metaphysical claims. Firstly, it should be noted that putting the word 'relative' or 'dynamic' or 'pragmatic' in front of the word 'a priori' changes the meaning of the latter completely. These kinds of a priori are not fixed, are not universal, are not necessary, and are not known by any sort of intuition, which means that they have none of the features of the traditional a priori. Nevertheless, for the purposes of discussing Reichenbach, Friedman, C. I. Lewis, Arthur Pap, etc. I generally maintain the term 'a priori' because these authors all use the term, even though I would prefer to give it up for the sake of clarity and replace it with 'constitutive elements'. That is, I would prefer to think of these philosophers as having theories of the role played by presuppositions underlying empirical inquiry in the physical sciences. Following the Kantian tradition, these presuppositions are constitutive of particular inquiries but they are not prior to experience or necessarily true. Rather they are deeply entrenched though fallible principles or elements that play a distinctive theoretical role in that they are preconditions of specific empirical theories; they must be assumed in order for the scientific investigations to be carried out. For example, calculus is necessary for Newtonian mechanics, as are Newton's laws of motion and universal gravity. Thus, even though I am discussing various views of the a priori, they are not really a priori at all in the traditional sense. Rather, there are various theories of constitutive elements in science; Kant's, in which the

---

[4] https://sites.google.com/site/socmetsci/what-is-the-metaphysics-of-science-1

constitutive elements are synthetic a priori, necessary and fixed, and twentieth-century views in which the constitutive elements are not fixed. So, we can understand the conceptual change in science as a change in the constitutive elements.

Secondly, my own starting point, my fundamental presupposition, as it were, is fallibilism. If we are consistently fallibilist, we will see that the search for universal principles – some fundamental axioms that are true for all time and for all places that can serve as a foundation – is a completely misguided project. Furthermore, we do not need absolutes anyway, because fallibilism does not imply that any belief is as justified as any other. There is a tremendous difference between saying that all knowledge is fallible (that we do not know anything for certain) and saying that we have no good evidence for our claims. We can rate our claims as justified or not, we can rank them in terms of margins of error that can be quantified with great precision, and we can give evidence for our claims. For example, the evidence gathered from a large double-blind study of the effectiveness of a medication is markedly stronger than a couple of random testimonials that some treatment "worked for me". Of course, as fallibilists emphasize, even the best evidence can turn out to be mistaken. Nevertheless, some evidence is better than other evidence, and we rely on these distinctions in making judgments. These are commonplaces and hardly bear repeating, except that philosophers tend to forget them.

Treating the fundamental constitutive elements in science as fallible leads immediately to the idea that they are hypotheses to be tested. Even if many of the topics covered by fallible hypotheses are the same as traditional metaphysics, our attitude towards our claims in these fields will be different. We are not claiming that we have discovered an absolute that is true in the actual or even in all possible worlds. Rather, we have made a claim to see where it leads and whether or not it can be developed in a way that remains consistent with empirical results. Despite my two caveats, there is nevertheless something of the rationalist element in science, whether it is considered to be a priori, metaphysical, or hypothetical. For example, Olivier Darrigol looks for rationalist arguments within science and makes a case that some elements of science are necessarily true, which would show that metaphysical claims have always been present in science as well as philosophy (Darrigol, 2014). Somewhat more broadly, there is the idea that physics needs philosophy, when it deals with fundamental issues of the nature of spacetime, or the nature of the universe in general, as found in a recent lecture by Carlo Rovelli at LSE.[5] The issue here is how to treat the elements of science that overlap with areas that have traditionally been called metaphysical. Part of my claim in my book is that

---

[5] https://youtu.be/IJ0uPkG-pr4

while they have been treated by many as a relative a priori in the twentieth century, they can best be seen as constitutive elements of science that are taken for granted in the beginning of inquiry and can be thought of as hypothetical and fallible.

7. Conclusion

I explain what the constitutive elements of scientific theories are and justify the claim that change in the constitutive elements in science is a model of radical conceptual change. Constitutive elements of scientific theories are integral to the foundations of scientific practice, and science cannot proceed without these elements in place. In this sense, the constitutive elements of science are taken for granted and are unquestioned. However, they can be questioned, given the proper context. The constitutive elements are fallible, just like all of the other elements of the scientific mosaic. Even though there is a long tradition of calling the constitutive elements of science a priori, the models of scientific change presented here are not about a priori knowledge, rather they are theories of the necessary preconditions that must be in place for a scientific practice to proceed. The literature on the relative a priori shows that these constitutive elements need not be permanently fixed. Nevertheless, taking them as changing need not lead to relativism. We do not need foundationalism or any kind of absolute to avoid relativism, but rather have theories that are fallible. Thoroughgoing fallibilism is not relativism, given that there can be independent constraints on our theories, both from other theories and directly from experience.

Bibliography

Arabatzis, T. & Kindi. V. (2008). The Problem of Conceptual Change in Philosophy and History of Science. In Vosniadou (Ed.) (2008), 345-373.

Ayer, A. J. (1950). *Language, Truth, and Logic.* 2nd ed. Dover.

Barseghyan, H. (2015). *The Laws of Scientific Change.* Springer.

Bradley, F. H. (1925). *Appearance and Reality: a Metaphysical Essay.* Macmillan.

Cassirer, E. (1923). *Substance and Function and Einstein's Theory of Relativity.* Open Court.

Chang, H. (2008). Contingent Transcendental Arguments for Metaphysical Principles. In Massimi (Ed.) (2008), 113-133.

Colodny, R. G. (Ed.) (1965). *Beyond the Edge of Certainty: Essays in Contemporary Science and Philosophy.* Prentice Hall.

Darrigol, O. (2014). *Physics and Necessity: Rationalist Pursuits from the Cartesian Past to the Quantum Present.* Oxford University Press.

Feigl, H. & Scriven, M. (Eds.) (1956). *Minnesota Studies in the Philosophy of Science.* University of Minnesota Press.

Friedman, M. (1997). Philosophical Naturalism. *Proceedings and Addresses of the American Philosophical Association*, 71(2), 5-21.

Friedman, M. (1999). *Reconsidering Logical Positivism.* Cambridge University Press.

Friedman, M. (2001). *Dynamics of Reason: the 1999 Kant Lectures at Stanford University.* CSLI Publications.

Friedman, M. (2008). Ernst Cassirer and Thomas Kuhn: The Neo-Kantian Tradition in History and Philosophy of Science. *Philosophical Forum*, 39(2), 239-252.

Garrison, J. (2019). The Myth that Dewey Accepts 'the Myth of the Given'. *Transactions of the Charles S. Peirce Society*, 55(3), 304-325.

Hacking, I. (1983). *Representing and Intervening.* Cambridge University Press.

Hacking, I. (1992). 'Style' for Historians and Philosophers. *Studies in History and Philosophy of Science*, 23, 1-20.

Hacking, I. (1999). *The Social Construction of What?* Harvard University Press.

Hacking, I. (2000). How Inevitable Are the Results of Successful Science? *Philosophy of Science*, 67, S58-S71.

Hanson, N. R. (1958). *Patterns of Discovery: An Inquiry into the Conceptual Foundations of Science.* Cambridge University Press.

Hanson, N. R. (1965). Newton's First Law: A Philosopher's Door into Natural Philosophy. In Colodny (Ed.) (1965), 6-28.

Herfeld, C. (2021). Understanding the Rationality Principle in Economics as a Functional A Priori Principle. *Synthese*, 198, 3329-3358.

Kuhn, T. S. (1962). *The Structure of Scientific Revolutions.* University of Chicago Press.

Kuhn, T. S. (2000). *The Road Since Structure.* University of Chicago Press.

Kusch, M. (Ed.) (2019). *The Routledge Handbook of Philosophy of Relativism.* Routledge.

Laudan, L. (1977). *Progress and its Problems: Towards a Theory of Scientific Growth.* University of California Press.

Laudan, L. (1987). Relativism, Naturalism and Reticulation. *Synthese*, 71, 221-234.

Levine, S. (2019). *Pragmatism, Objectivity, and Experience.* Cambridge University Press.

Lewis, C. I. (1923). Pragmatic Conception of the A Priori. *The Journal of Philosophy*, 20(7), 169-177.

Luchetti, M. (2021). Constitutive Elements in Science Beyond Physics: The Case of the Hardy–Weinberg Principle. *Synthese*, 198, 3437-3461.

Massimi, M. (Ed.) (2008). *Kant and Philosophy of Science Today.* Cambridge University Press.

Nersessian, N. J. (1989). Conceptual Change in Science and in Science Education. *Synthese*, 80(1), 163-183.

Nersessian, N. J. (2003). Kuhn, Conceptual Change, and Cognitive Science. In Nickles (Ed.) (2003), 179-211.

Nersessian, N. J. (2008). *Creating Scientific Concepts.* MIT Press.

Nickles, T. (Ed.) (1980). *Scientific Discovery, Logic and Rationality.* D. Reidel,

Nickles, T. (Ed.) (2003). *Thomas Kuhn.* Cambridge University Press.

Pap, A. (1946). *The A Priori in Physical Theory.* King's Crown Press.

Parrini, P. (1983). *Empirismo logico e convenzionalismo: saggio di storia della filosofia della scienza.* Franco Angeli.

Parrini, P. (1998). *Knowledge and Reality: An Essay in Positive Philosophy.* Kluwer Academic Publishers.

Reichenbach, H. (1965). *The Theory of Relativity and A Priori Knowledge.* University of California Press.

Sankey, H. (2019). The Relativistic Legacy of Kuhn and Feyerabend. In Kusch (Ed.) (2019), 379-387.

Sellars, W. (1956). Empiricism and the Philosophy of Mind. In Feigl & Scriven (Eds.) (1956), 253-329.

Shapere, D. (1980). The Character of Scientific Change. In Nickles (Ed.) (1980), 648-649.

Shapere, D. (1984). Objectivity, Rationality, and Scientific Change. *Proceedings of the Biennial Meeting of the Philosophy of Science Association*, 1984, 637-663.

Shaw, J. (2020). The Search for Kuhn-Loss. Academia. Retrieved from: https://www.academia.edu/29796270/The_Search_for_Kuhn_Loss.

Soler, L., Sankey, H., & Hoyningen-Huene, P. (Eds.) (2008). *Rethinking Scientific Change and Theory Comparison.* Springer.

Stump, D. J. (1991). Fallibilism, Naturalism and the Traditional Requirements for Knowledge. *Studies in History and Philosophy of Science*, 22, 451-469.

Stump, D. J. (2015). *Conceptual Change and the Philosophy of Science: Alternative Theories of the A Priori.* Routledge.

Stump, D. J. (2021) Fallibilism versus Relativism in the Philosophy of Science *Journal for General Philosophy of Science.* https://doi.org/10.1007/s10838-021-09579-x.

Toulmin, S. (1961). *Foresight and Understanding: An Enquiry into the Aims of Science.* Indiana University Press.

Tsou, J. (2010). Putnam's Account of Apriority and Scientific Change: Its Historical and Contemporary Interest. *Synthese*, 176(3), 429-445.

Uebel, T. (2012). De-Synthesizing the Relative A Priori. *Studies in History and Philosophy of Science Part A*, 43(1), 7-17.

Vosniadou, S. (Ed.) (2008). *International Handbook of Research on Conceptual Change.* Routledge.

Walter, S.A. et al. (Eds.) (n.d.). *Henri Poincaré Papers*, Doc. 3-18-2. Retrieved from: http://henripoincarepapers.univ-nantes.fr.

Index

T

Lightning Source UK Ltd.
Milton Keynes UK
UKHW022355120122
397039UK00002B/5/J

9 781648 892950